KUHN'S EVOLUTIONARY SOCIAL EPISTEMOLOGY

Kuhn's *Structure of Scientific Revolutions* (1962) has been enduringly influential in philosophy of science, challenging many common presuppositions about the nature of science and the growth of scientific knowledge. However, philosophers have misunderstood Kuhn's view, treating him as a relativist or social constructionist. In this book, Brad Wray argues that Kuhn provides a useful framework for developing an epistemology of science that takes account of the constructive role that social factors play in scientific inquiry. He examines the core concepts of *Structure* and explains the main characteristics of both Kuhn's evolutionary epistemology and his social epistemology, relating *Structure* to Kuhn's developed view presented in his later writings. The discussion includes analyses of the Copernican revolution in astronomy and the plate tectonics revolution in geology. The book will be useful for scholars working in science studies, sociologists, and historians of science, as well as philosophers of science.

K. BRAD WRAY is an associate professor of philosophy at the State University of New York, Oswego. He has published extensively on the epistemology of science, Kuhn's philosophy of science, and the anti-realism/realism debate. He was the guest editor of a special issue of the journal *Episteme*, on the theme of Collective Knowledge and Science, and he is also the editor of an epistemology textbook, *Knowledge and Inquiry* (2002).

T0175612

KUHN'S EVOLUTIONARY SOCIAL EPISTEMOLOGY

K. BRAD WRAY

State University of New York, Oswego

CAMBRIDGE UNIVERSITY PRESS
Cambridge, New York, Melbourne, Madrid, Cape Town,
Singapore, São Paulo, Delhi, Tokyo, Mexico City

Cambridge University Press
The Edinburgh Building, Cambridge CB2 8RU, UK

Published in the United States of America by
Cambridge University Press, New York

www.cambridge.org
Information on this title: www.cambridge.org/9781107012233

First published 2011

A catalogue record for this publication is available from the British Library

Library of Congress Cataloguing in Publication data
Wray, K. Brad, 1963–
Kuhn's evolutionary social epistemology / K. Brad Wray.
p.   cm.
Includes bibliographical references and index.
ISBN 978-1-107-01223-3 (hardback)
1. Science–Philosophy.   2. Knowledge, Theory of.   3. Kuhn, Thomas S.
4. Social epistemology.   I. Title.
Q175.W78   2011
501–dc23
2011026075

ISBN 978-1-107-01223-3 Hardback

For Lori

Contents

Figures and table

Figures

Table

Acknowledgements

I began working on Kuhn's epistemology in 2001. After a series of publications, and with a sabbatical leave approaching, I began to think about writing a book on Kuhn's epistemology of science, one that would take account of his later work, much of it published in *The Road since Structure*. I believed that this work was largely and unfortunately neglected by philosophers, and that a fuller picture of his view was available to those who attended to it. I also believed that there were still many important insights that philosophers of science could gain from his work as we seek to develop an epistemology of science.

My interest in Kuhn's work, though, extends back further to my time as a graduate student at the University of Western Ontario. I was fortunate enough to study Kuhn in a directed reading with my thesis supervisor, John Nicholas. And my understanding of Kuhn's views has been enriched from teaching Kuhn's work. In my efforts to present Kuhn's views to my students over the years I have been able to discover common sources of resistance to and misunderstandings of his work.

The secondary literature on Kuhn is vast, and necessarily I have had to be selective in whose work I discuss. In writing the book, though, I have benefited from a number of Kuhn scholars. I have found the most useful to be the following: Ernan McMullin, Ian Hacking, Larry Laudan, Paul Hoyningen-Huene, Hanne Andersen, and Alexander Bird. These scholars have offered both valuable insights into understanding Kuhn's views and interesting interpretations and criticisms that warrant serious consideration. Though I disagree with each of them on some point or other, their research has helped me clarify my own thoughts on Kuhn's view.

Once I set about writing the book, I relied on the generosity of many people, who, in one way or another, helped me complete the project.

I presented various papers on Kuhn at a variety of conferences, including the following: the Canadian Society for History and Philosophy of

Science, the Society for the Social Studies of Science, the Philosophy of Science Conference in Dubrovnik, the American Philosophical Association, the International Congress of Logic, Methodology, and Philosophy of Science, the Science Studies Research Group at Cornell University, and a workshop on Relativism, Philosophy of Science, and Social Studies of Science at the Helsinki School of Economics. These were very useful sources of feedback as I worked on my research on Kuhn. In addition, I have also presented numerous papers on Kuhn at department colloquia to my supportive colleagues at the State University of New York at Oswego. Financial support from the Office of International Education at SUNY-Oswego, the Dean of the College of Liberal Arts and Sciences at SUNY-Oswego, and United University Professions, my union, helped me to cover the travel costs to the various conferences at which I presented my work on Kuhn.

A crucial turning point in this project was my sabbatical in the 2008/09 academic year. The fall semester of my sabbatical leave was spent as a Visiting Scholar in the Department of Science and Technology Studies at Cornell University. This department is a collection of scholars whose training is mostly in sociology and history of science. This was a formative experience as I worked on the book. More than ever before, I saw the differences between the ways sociologists and historians approach the study of science and the ways philosophers do. Discussions and exchanges with Peter Dear and Trevor Pinch were especially helpful in this regard. I saw the need to make clear to historians and sociologists how it is that philosophers see science. In the book, I attempt to make clear where the key disagreements are between philosophers and sociologists of science in an effort to move beyond the current rift between scholars in the two fields. I thank Michael Lynch for hosting me during my visit at Cornell by agreeing to be my sponsor. Michael has been a source of encouragement for a number of years. Most importantly, as editor of *Social Studies of Science*, he published my research on Kuhn. I also thank Peter Dear for allowing me to attend his class on the history of science and participate in his seminar on the historiography of science.

In the spring term of my sabbatical leave I worked mainly from home, but spent a wonderful week in Finland, participating in a workshop on relativism, philosophy of science, and science studies organized by Kristina Rolin, at Aalto University, which was then called the Helsinki School of Economics. Critical feedback and encouragement from both Kristina and Martin Kusch was very helpful.

A number of people read either the whole manuscript or large segments of it as I worked on it, including Kristina Rolin, Leigh Bacher, Hanne Andersen, an anonymous referee, David Hull, and my partner, Lori Nash.

I have been sharing my work with Kristina Rolin since I met her in 1999, at the Logic, Methodology, and Philosophy of Science Conference in Krakow, Poland. I have benefited greatly from our overlapping interest in the social epistemology of science, and she has provided constructive feedback on most of the chapters of the manuscript.

Leigh Bacher, my colleague in Psychology at SUNY-Oswego, read portions of the book. Her feedback was also very useful. I have also been collaborating with Leigh on a project examining how college students learn scientific reasoning skills, a project that has provided us with numerous opportunities to share insights about how scientific research is done.

Hanne Andersen read the complete manuscript as a referee for Cambridge University Press, providing valuable recommendations for improving it for publication. Similarly, the second, anonymous reader for the Press also provided numerous valuable suggestions that have substantially improved the book.

David Hull read the complete manuscript. It is unfortunate that he passed away before the book made it into print. David has been a mentor for me, guiding me in my pursuit of a career in the philosophy of science. His work in the epistemology of science has profoundly shaped my own work, which is evident throughout the book. I also thank Marc Ereshefsky, at the University of Calgary, for introducing me to David and his work in the late 1990s.

My partner, Lori Nash, read and reread the manuscript, once completing it in a two-day sitting. She has been a wonderful support throughout my career, encouraging me to clarify my arguments and to pursue my dreams. She has made my life exciting and fun-filled. I thank her for the continuous encouragement and the wonderful times together.

I thank the publishers and editors of the following journals for permission to include portions of previously published papers that have been included in the book:

Wray, K. B. (forthcoming: 2011). "Kuhn and the Discovery of Paradigms," *Philosophy of the Social Sciences*, published by Sage;

2010. "Kuhn's Constructionism," *Perspectives on Science: Historical, Philosophical, Social*, 18:3, 311–27, published by MIT Press;

2007. "Kuhnian Revolutions Revisited," *Synthese: An International Journal for Epistemology, Methodology, and Philosophy of Science*, 158:1, 61–73, published by Springer;

2005. "Rethinking Scientific Specialization," *Social Studies of Science: An International Review of Research in the Social Dimensions of Science and Technology*, 35:1, 151–64, published by Sage;

2005. "Does Science Have a Moving Target?," *American Philosophical Quarterly*, 42:1, 47–58, published by University of Illinois Press;

2003. "Is Science Really a Young Man's Game?," *Social Studies of Science: An International Review of Research in the Social Dimensions of Science and Technology*, 33:1, 137–149, published by Sage.

Finally, I thank Hilary Gaskin, Senior Commissioning Editor, Philosophy, and Anna Lowe, Assistant Editor, Humanities, at Cambridge University Press for overseeing the project. Their encouragement and assistance have been greatly appreciated. In addition, I thank Thomas O'Reilly, Dr. Matthew Davies and Christopher Feeney for their assistance in preparing the book for publication.

Introduction: Kuhn's insight

In *The Structure of Scientific Revolutions* Kuhn developed a novel and interesting account of the dynamics of scientific change, one that was deeply at odds with the assumptions that had previously informed the outlook of philosophers of science. To many of his readers it seemed that whenever Kuhn denied a widely accepted philosophical assumption about science, he offered a paradox in its place.

To begin with, Kuhn alleged that scientific knowledge was not cumulative. He is famous for drawing our attention to what has come to be called "Kuhn-loss," the "knowledge" allegedly lost when one theory replaces another. Yet he adamantly insisted that there is scientific progress.

He also claimed that observational data could not provide a foundation for scientific knowledge. Instead, he insisted that data are pliable and thus scientists could not unequivocally settle disputes by appealing to data. Yet he emphasized the importance of scientists' work on relatively small, manageable, esoteric problems, which seemed to treat data as capable of disclosing unequivocal answers to the questions driving research. These problems he called the puzzles of normal science. Indeed, in the context of normal science, as Kuhn describes things, the data seem to have an almost veto power. Rather than posing a threat to the theory assumed in research, discrepancies between expectations and results show the incompetence of the scientist.

He also claimed that scientists were not especially open-minded or critical, as Karl Popper claimed. In fact, Kuhn claimed that scientists are remarkably uncritical with respect to the accepted theories. Further, he suggested that the education of scientists was dogmatic, never inviting the student to question the accepted theory. And scientific inquiry, he claimed, was tradition-bound.

No wonder *Structure* was met with fierce criticism. Kuhn was giving us an account of science very different from the positivists' account. It seemed that he was denying every assumption that the positivists made about science.

Kuhn was not alone in challenging the received view. He was part of a new wave in philosophy of science, the historical school. Members of this school believed that philosophers could benefit greatly from examining the history of science. A study of the history of science, they thought, would disclose the way in which scientific inquiry *really* worked. The historical school did not question the epistemic authority of science and scientists. Rather, those who subscribed to this view sought to understand science as it was really practiced. They were not interested in a rational reconstruction or idealization of science.

Initially, the work of the historical school was greeted with enthusiasm, as Kuhn and others working in a similar vein drew attention to the discrepancies between the idealized picture of science that philosophers had been working with and the real world of science, as gleaned from an examination of the history of science. He and the others, drawing on the history of science, promised to enrich our understanding of science.

In developing his own view of science, Kuhn invoked a variety of engaging metaphors that seemed to underscore the inadequacy of traditional philosophical accounts of science. Changes of theory were described as scientific *revolutions*, comparable, in some respects, to political revolutions. They were very unsettling events that required radical breaks with the past. Scientists involved in such events were alleged to undergo something like a conversion experience, much like a religious conversion, an experience that seemed to admit of no rational defense. And, science moved from one paradigm to another. Kuhn likened this move to a gestalt shift, thus raising questions about the relationship between the world and our theories. Kuhn even compared the change scientists underwent when they accepted a new theory to a world change. That is, after a scientific revolution, scientists not only work with a new theory, they seem to work in a *new world*. These metaphors and comparisons were deeply unsettling to many philosophers, even if at the same time they were liberating and promised to offer new insights into science.

The publication of *Structure* quite quickly altered both the philosophy of science and the sociology of science profoundly. Both fields were set in new directions.

Sociology of science, and its successor project, science studies, became more involved in investigating the cognitive dimensions of science than ever before. Prior to the publication of *Structure*, sociologists of science studied the institutional structure of science and the impact of external factors on science, like developments in commerce and trade. After *Structure*, though, sociologists started to examine how social factors affected the outcome of scientific disputes, determining the way in which disputes were resolved. These sorts of investigations were perceived by most philosophers as unwelcome and threatening intrusions into the traditional domain of philosophy. And they were met with fierce resistance. Because Kuhn was regarded by many sociologists as a source of inspiration, many philosophers held him responsible for encouraging these new developments in the sociology of science.

Even before these developments in the sociology of science, however, philosophers of science were critical of Kuhn's work. The tone was set early by Popper and his fellow Popperians, when Popper and Kuhn engaged each other at the 1965 London International Colloquium in the Philosophy of Science. The Popperians were most disturbed by Kuhn's account of normal science. Popper had emphasized the critical attitude of science, the readiness to subject any belief to empirical testing. Kuhn, on the other hand, described the normal research activities of scientists as dogmatic. Scientists, according to Kuhn, looked at the world uncritically, unreflectively employing the concepts of accepted theories. Moreover, science education was described as a process that made scientists myopic, often even unable to see evidence contrary to their theoretical expectations. Indeed, one might wonder how on Kuhn's account a change of theory was even possible.

This dimension of Kuhn's view, in combination with his unsettling remarks about the apparently non-rational process that leads to a change in theory, led to the development of a very negative reading of Kuhn. According to this reading, Kuhn's account of science and scientific change threaten the rationality of science. If scientists really are in the grip of the accepted theory to the extent that Kuhn implies, and it takes something like a religious conversion to set them free, it is hard to see how theory change could be a rational process. Many found it very difficult to reconcile Kuhn's picture of science with the accepted view of science as critical inquiry, an enemy of dogmatism, and driven by a healthy, skeptical attitude. Consequently, many thought that Kuhn's account of science was deeply mistaken.

It was not only Kuhn who was mistaken. A generation of sociologists was under his spell, extending his ideas in ways that even Kuhn found distressing. Indeed, by the mid 1980s, Larry Laudan felt the need to write a book aimed at saving us from the Kuhnians, showing why their view of science is deeply mistaken (see Laudan 1984). By the time Laudan published his book, though, it was not Kuhn's own view that was the real object of concern. Rather, it was a particular reading of Kuhn, one influenced substantially by developments in the sociology of science.

In much of his later work Kuhn tried to correct some of the misunderstandings of his position. In his attempts to defend his views from criticism, Kuhn tried to clarify his account, modifying and developing it along the way. But unfortunately many of these developments went unnoticed by his critics and commentators. To a large extent, philosophers of science seemed content to accept the existence of a standard Kuhnian position, a threatening but ultimately indefensible position against which they would define their own positions. This attitude, I believe, is quite unfortunate as Kuhn's developed position is thoughtful, offering important insights into the nature of scientific change and scientific knowledge. Indeed, Kuhn also offers important insight into how we should study science as philosophers.

My aim in this book is to make a case for taking Kuhn's developed view seriously. Kuhn offers us a framework for developing an epistemology of science. Given the social nature of scientific inquiry, Kuhn believed that an epistemology of science needs to be a *social* epistemology. He also believed that an epistemology of science needs to be an *evolutionary* epistemology. Both "social epistemology" and "evolutionary epistemology" are labels that pick out a wide range of projects. I aim to clarify the nature of Kuhn's approach to epistemology, outlining the respects in which it is an evolutionary epistemology and those in which it is a social epistemology.

None of the existing books about Kuhn's philosophy of science give adequate attention to the social dimensions of scientific inquiry. Nor have they given much attention to research in the sociology of science. Such research, I argue, is extremely relevant to advancing the goals of epistemologists of science. Moreover, none of the existing books take account of Kuhn's attempt to develop an evolutionary epistemology. My aim is to address these shortcomings.

It should be noted that the continuity between Kuhn's later work and the view he developed in *Structure* is quite extensive. Thus, though in developing his view Kuhn revised his views in significant ways, he was motivated, to a large extent, by the desire to clarify what he was trying to say in the early 1960s. Indeed, some of the developments in his later work are best described as extensions of the project that he began with *Structure*.

OVERVIEW

My aims in this book are: (1) to clarify the nature of Kuhn's epistemology of science, (2) to offer a defense of his epistemology, and (3) to clarify the relationship between Kuhn's views and recent work in sociology of science and science studies. Kuhn's view is too often mistakenly characterized as an unacceptable form of constructionism or relativism. Motivating my study is a concern to show that Kuhn has a positive legacy to offer philosophers of science, a constructive and insightful framework for developing an epistemology of science. Moreover, I aim to show that philosophers cannot afford to be dismissive about sociology of science. Given the social nature of scientific inquiry, sociological studies of science will play a key role in developing an adequate descriptive account of scientific inquiry and change.

Kuhn continued to develop his epistemology of science until the end of his life. Many of the later developments in his view, however, have been neglected by philosophers, who have tended to focus on the view articulated in *Structure*. This is unfortunate, as his developed view clarifies the nature of revolutionary changes in theory, one of the most contentious parts of his position as presented in *Structure*. Most significantly, he replaces the highly criticized notion of a paradigm change with the notion of a taxonomic or lexical change. I aim to show how such changes are both radical and yet rationally defensible.

Further, Kuhn develops an account of the process that leads to the creation of new scientific specialties, a topic that has been largely neglected by philosophers, though discussed extensively by sociologists and historians of science. Central to Kuhn's account of specialization is a radical understanding of the end or goal of scientific inquiry. Traditionally, philosophers have uncritically assumed that truth is the end of inquiry, and the success of science is best explained in terms of the pursuit of this goal. Kuhn, on the other hand, suggests that science is better conceived as developing through a process of increasing specialization. This dimension of Kuhn's project has been largely overlooked, in large part because he never presented his views on specialization systematically. Moreover, specialization has typically not been a topic of concern to philosophers of science. I aim to provide a clear and systematic presentation of Kuhn's account of specialization. Further, I aim to articulate the philosophical relevance of Kuhn's account of scientific specialization, showing how the process of specialty formation is driven by cognitive or epistemic considerations. In this respect, Kuhn's account of specialization differs significantly from sociological accounts, which tend to emphasize the social dimensions of the change, and downplay the epistemic dimensions. In fact, Kuhn came to believe that specialization is one of the means by which scientists are able to develop an increasingly accurate and comprehensive understanding of the world.

Unlike some other philosophers writing on Kuhn, I aim to critically analyze the relation between Kuhn's view and sociological studies of science. *Structure* had a profound impact on the sociology of science. But the directions in which sociology of science developed has created a rift between philosophy of science and sociology of science, and Kuhn is often thought to be partly responsible for this state of affairs. On the one hand, I aim to show how Kuhn's view differs from many of the sociological studies of science that were inspired by his work. Consequently, I argue, Kuhn has been unfairly criticized as a social constructionist. I believe that Kuhn is nonetheless a constructionist of sorts, though we need to take some care in distinguishing the form of constructionism he endorses from other untenable forms. On the other hand, I aim to show that given Kuhn's conception of the epistemology of science, and especially his view that the loci of theory changes are research communities, philosophers will have to either work with sociologists of science or draw on research in the sociology of science. This will enable philosophers to develop a richer descriptive account of scientific change. It is unfortunate that Kuhn never systematically articulated the relationship between his view and the views of contemporary relativist sociologists of science.

This book is in three parts.

The first part is titled "Revolutions, paradigms, and incommensurability." In it I re-examine some of the most important and contentious concepts that Kuhn employed in *Structure* with the aim of clarifying how his view developed with respect to these concepts. Though now widely used in philosophy of science, these concepts are often used in ways very different from the ways in which Kuhn used them or intended them to be used.

I begin with Kuhn's modified account of scientific revolutions, developed in the later part of his career. Originally, in *Structure*, Kuhn characterized scientific revolutions as paradigm changes. But because of the variety of meanings "paradigm" had in *Structure*, the notion of a paradigm change led to many misunderstandings and much criticism. Later, in an effort to correct misunderstandings and address his critics, Kuhn came to characterize scientific revolutions as involving taxonomic or lexical changes, a reordering of the relationships between concepts in a theory. I defend Kuhn's revised account of scientific revolutions against a series of common criticisms.

I also examine in detail the Copernican revolution in early modern astronomy to illustrate the explanatory power of Kuhn's account. Kuhn's own book-length treatment of this episode in the history of science was published before he published *Structure*, and thus before he had worked out the details of his account of scientific change. This episode in the history of science has also been the subject of much debate, and the historical scholarship on the topic has developed extensively since Kuhn published *The Copernican Revolution*. Consequently, it is worth re-examining this episode in the history of science with the aid of Kuhn's developed account of scientific change.

In developing his view on revolutions, Kuhn did not completely discard the notion of a paradigm. As a result, I will be clarifying the role that paradigms play in Kuhn's developed philosophy of science. Because theory change is no longer characterized as paradigm change, one might be led to think that paradigms have little significance in his developed view. This is not so. Paradigms still function as the widely recognized concrete scientific achievements that are used as models for solving hitherto unsolved problems in a field. They are also the means by which young aspiring scientists learn the norms, standards, practices, concepts, and theories in their field. Hence, paradigms play an essential role in the socialization of young scientists. Further, I argue that Kuhn's discovery of the concept "paradigm" exemplifies the complex process of discovery in science. Hence, as odd as it may sound, by the time he wrote *Structure*, Kuhn had not yet discovered what a paradigm was.

Part I ends with an examination of the role of incommensurability in science. I distinguish the various ways in which Kuhn used the term "incommensurable" and identify the epistemic significance of each type of incommensurability. Incommensurability is often thought to pose a significant threat to the rationality of theory change. If two theories are not even comparable, it is difficult to understand how scientists are able to reach a rational judgment about which of the theories is superior from an epistemic point of view. Initially, in *Structure*, Kuhn appealed to the concept of incommensurability in order to capture the fact that scientists lack a common measure by which to evaluate competing theories. This is why revolutionary changes can be such protracted affairs. But in his efforts to address his critics, Kuhn talked more and more about "meaning incommensurability," the fact that a single term, like "mass" for example, has a different meaning in competing theories. Kuhn also came to describe the lexicons of neighboring fields as incommensurable. In fact, he came to believe that the incommensurability of the lexicons of neighboring scientific specialties plays an important role in isolating scientists, and thus allowing them to develop concepts appropriate to the phenomena they study. Although meaning-incommensurability has attracted the most attention from philosophers writing about incommensurability, I believe that it has less epistemic significance than the two other forms of incommensurability described here.

The second part of this book is titled "Kuhn's evolutionary epistemology." In it I examine the aspects of Kuhn's epistemology of science that make it an evolutionary epistemology.

The popularity of and enthusiasm for evolutionary epistemologies has waxed and waned over the last five decades. And there is hardly a uniform understanding about what makes an epistemology an evolutionary epistemology. I aim to clarify the senses in which Kuhn's epistemology of science is aptly described as an evolutionary epistemology. In addition, I aim to show that his evolutionary perspective on science is an important resource for developing an adequate epistemology of science. His evolutionary perspective, though, profoundly alters the way we see science. Indeed, I believe it is the magnitude of the changes caused by this change in perspective that has led to so many misunderstandings of Kuhn's view.

Kuhn is widely recognized as one of the pioneers of the historical school in philosophy of science, a group that aimed to look to the history of science as a source of data for developing a philosophy of science. Such an approach to the study of science was meant to lead to a more accurate account of science, in contrast to the idealizations that emerged from the rational reconstructions of his predecessors. But Kuhn changed his mind about the relevance of the history of science to the philosophy of science. He came to believe that the key lesson philosophers must learn from history is a particular perspective, a developmental or historical or evolutionary perspective.

Traditionally, philosophers have assumed that science aims for the truth, that is, to mirror a reality that is indifferent and essentially unchanging. Moreover, traditionally, it is assumed that the history of science is marked by a steady accumulation of knowledge, often aided by the development of unifying theories, theories that bring together disparate phenomena under a set of laws. Kuhn challenged this traditional picture of science in a variety of ways. First, in *Structure* Kuhn suggested that science is best seen as moved from behind, rather than aiming at some goal set by nature in advance. He compares scientific change to evolutionary change by natural selection. According to Darwin, the process of biological change is not teleological. This was Darwin's most radical innovation. Similarly, Kuhn maintains that science is not aptly described as moving toward a fixed goal, set by nature in advance. Instead, scientists are moved by research agendas set by their predecessors, and they work with instruments and theories developed by their predecessors. I defend Kuhn's view. I argue, in addition, that scientists must even determine what phenomena their theories aim to account for. In this respect, the target at which scientists aim in their efforts to develop theories is not predetermined.

Once we adopt the developmental perspective that Kuhn recommends, we realize that scientists are always working within research traditions, working from sets of beliefs inherited from their predecessors. Moreover, their evaluations of theories are comparative, for they are unable to compare their theories directly with a mind-independent reality. Further, the increasing predictive accuracy achieved in mature fields is not to be explained by citing the (alleged) fact that we are getting increasingly closer to the truth. Rather, our success in science is better explained as a consequence of the increasing specialization in science. As new specialties are formed, scientists can develop instruments, practices, and concepts suited to a narrower range of phenomena. The result is an increase in predictive power.

I also compare Kuhn's account of specialty formation with sociological and historical accounts of the process. The accounts of specialization developed by historians and sociologists of science tend to privilege the social dimension of the change that occurs, and treat the conceptual changes as derivative. Kuhn, on the other hand, gives a privileged place to the conceptual dimension of the developments of a new specialty. Given Kuhn's account of specialization, the process that leads to the creation of new specialties is of great importance to philosophers of science interested in the epistemic dimensions of science. I examine two case studies to illustrate Kuhn's account: the formation of endocrinology as a field and the formation of virology as a field of research.

Kuhn's account of specialization is important for three reasons: (1) it provides insight into the often overlooked cognitive or epistemic dimensions of the process; (2) it makes clear why specialization is relevant to philosophers of science, and not just sociologists and historians of science; and (3) it is an aspect of Kuhn's developed account of scientific change that is still either largely neglected or misunderstood. Specialization, I argue, will prove to be an important topic in developing a richer understanding of both scientific inquiry and scientific knowledge.

The third and final part of the book is titled "Kuhn's social epistemology." Here I examine the respects in which Kuhn's epistemology of science is aptly described as a social epistemology. I also provide some direction as to what we need to do next as we seek to develop a Kuhnian social epistemology of science.

I begin by examining the charge that Kuhn is a social constructionist. Though sociologists of science often enthusiastically accept such a label, indeed, even self-consciously describe themselves as constructionists, philosophers of science are averse to being called constructionists. In philosophical circles the term still carries connotations of relativism and irrationalism. In addition, for philosophers of science constructionism also connotes a commitment to externalism and nominalism. Further, because the label "constructionist" is used in a variety of ways, it is far from clear what is meant when someone is labeled a constructionist.

I aim to clarify the relationship between Kuhn's epistemology and constructionism. Contrary to what some of Kuhn's critics claim, I argue that Kuhn is an internalist, believing that changes in theory are ultimately caused by a consideration of epistemic factors, not external factors. Kuhn does in fact attribute a significant role to subjective factors in theory change, arguing that such factors are responsible for ensuring that there is an efficient division of labor and competing theories are developed. It is only when competing theories are developed that the epistemic merits and shortcomings of the theories emerge. And only when the epistemic merits and shortcomings of competing theories are revealed can a rational choice be made between competing theories. I also argue that Kuhn is not a radical nominalist. Kuhn does not believe that there are *no* constraints imposed by the world on how a successful theory groups things in the world. In fact, he is quite insistent that the mind-independent world imposes constraints that are irreconcilable with some hypotheses. Still, contrary to what many philosophers and scientists claim, Kuhn does not believe that there is a single ultimate way our theories need to group things. My analysis of Kuhn's constructionism is meant to clarify the relationship between his view and popular contemporary sociological views of science, with special attention to his relationship to the views of the Strong Programme.

I then examine the ways in which Kuhn's epistemology of science is aptly described as a social epistemology. Most importantly, Kuhn regards the research community or specialty as the locus of change in science. A change of theory is not effected merely by a change in the view accepted by a scientist. Rather, a change of theory is a change in the research community. Thus for Kuhn theory change is a form of social change. This is evident from the way Kuhn characterizes the development of a mature field, from a stage of normal science, to a crisis, to a revolution resulting in a new phase of normal science. Consequently, in our efforts to better understand science and scientific change we need to develop a better understanding of the type of social change that is a theory change. Moving forward on this issue will require philosophers to work with, or at least draw on, the work of social scientists.

Finally, I examine the issue of how a new theory comes to be accepted in a research community, that is, how a new theory is able to replace a long-accepted theory. My aim is, in part, to demonstrate how empirical research on the social dimensions of science can shed light on both the dynamics of theory change and the role evidence plays in scientific disputes. My point of departure is two claims Kuhn makes about the role of young and old scientists in episodes of theory change. Kuhn claims that (1) young scientists are more likely than older scientists to create new theories, for young scientists have less invested in the accepted theory than older scientists. Young scientists, he suggests, are also less under the spell of the accepted theory, and thus better able to see the world differently from how they have been taught to see it by the accepted theory. Further, Kuhn claims that (2) older scientists are especially resistant to theory change. Because they have long worked within the tradition defined by the accepted theory, and often contributed to its development, older scientists are reticent to *accept* a new theory. To do so, is, to some extent, to undermine their life's work. I raise concerns about both of Kuhn's claims about age and theory change.

I then examine a particular case of theory replacement, the acceptance of the theory of plate tectonics in geology in the 1960s. This example provides me with an opportunity to demonstrate what philosophers of science can learn from empirical studies of science, the sorts of studies that sociologists of science conduct. Drawing on an existing empirical study of this episode in the history of science, I show that different scientists responded to the new theory in different ways. Some geologists accepted the new theory even before the bulk of the evidence supported it. Others were moved by the new evidence gathered in the 1960s. And still others were late to accept the new theory. The range of responses to the evidence, I argue, need not threaten the rationality of science. In fact, the range of responses suggests that the research community was well structured in certain respects. Early accepters ensured that the research community as a whole was not too dogmatic and close minded. And late accepters ensured that it was not too fickle.

Thus, we still stand to learn much from Kuhn's philosophy of science. Indeed, I argue that both his descriptive account of science and his orientation to the study of science, the evolutionary perspective, are the keys to developing a richer understanding of science and scientific knowledge. Hence, Kuhn's descriptive account of science and his evolutionary perspective provide a framework for developing a general philosophy of science.

This book is, first and foremost, concerned with philosophy of science. It is about the epistemology of science. But because I explore the relationship between Kuhn's views and work in the sociology of science, the book also provides sociologists with insight into the source of the tensions between sociologists and philosophers of science. Much of the animosity between scholars in the two fields is due, I believe, to misunderstandings about the other field. And some of the misunderstandings are due to the fact that sociologists and philosophers employ different concepts, and thus, to some extent, talk past each other. Overcoming such communication barriers is genuinely challenging. But I do not think that philosophers can afford to neglect the work of sociologists of science. A viable epistemology of science needs to draw on work in the sociology of science.

PART I

Revolutions, paradigms, and incommensurability

CHAPTER I

Scientific revolutions as lexical changes

There is no better place to begin a study of Kuhn's developed epistemology of science than with his remarks on scientific revolutions. This is so for three reasons. First, revolutions figure so importantly in Kuhn's account of scientific change. It is their structure that he was trying to elucidate in *The Structure of Scientific Revolutions*. And it is their existence which supports his non-cumulative account of scientific knowledge. Second, revolutionary scientific changes were the focus of much of the criticism against Kuhn's account of science. According to Kuhn, the development of scientific knowledge is punctuated by scientific revolutions, dramatic and unsettling events that undermine the traditional picture of the growth of scientific knowledge as cumulative. Such an account of science was widely perceived as posing a significant threat to the rationality of science. Third, in his later work Kuhn develops a new definition and understanding of scientific revolutions, one designed to avoid the pitfalls of his earlier characterization of scientific revolutions as paradigm changes.

In this chapter, I examine Kuhn's developed account of scientific revolutions. He no longer identifies revolutions as paradigm changes. Rather, a revolution involves the replacement of an accepted scientific lexicon or taxonomy with a new one. Such changes are precipitated by crisis in the research community. And the resolution of the dispute between advocates of the competing theories or lexicons cannot be resolved by means of shared standards. Importantly, Kuhn regards the research community, or scientific specialty, as the locus of theory change and scientific change in general. Revolutions are not just changes in individual scientists' beliefs. This helps us understand why Kuhn stopped comparing revolutionary changes to gestalt shifts. Research communities are incapable of experiencing gestalt shifts. Hence, a revolutionary change occurs only when a *research community* replaces the theory with which it works with another theory. This is one important respect in which Kuhn's epistemology of science is aptly described as a *social* epistemology.

Historians and philosophers of science have raised concerns about the concept of a "scientific revolution." Some historians, Steven Shapin (1996) and Betty Jo Teeter Dobbs (2000), for example, suggest that "*the* Scientific Revolution" of the early modern era was merely a rhetorical construction. They claim that though the scientific revolution has figured prominently as an organizing idea in the discipline of history of science, it has now lost its utility.

The concerns of philosophers are somewhat different. Philosophers have expressed concern about the fact that revolutionary changes in science, if they in fact occur, *may not* be rationally defended (see, for example, Chen and Barker 2000 and van Fraassen 2002, lecture 3). Thus, some have expressed the fear that if Kuhn's account of the process of theory change is correct, theory change cannot possibly be rational (see Lakatos 1970/1972 and Laudan 1984). This has led some philosophers to seek means to mitigate this threat to the rationality of science by either denying the existence of revolutionary changes or domesticating them in some way (see Laudan 1984 and Andersen *et al.* 2006).

My aim is to examine the extent to which Kuhn's political metaphor, his comparison of theory change in science to radical political change, is appropriate. This metaphor may seem to support his critics' reading, according to which Kuhn regarded revolutionary changes as irrational and resolved by means of political power. But I aim to show that there is a more charitable and insightful reading, one that matches Kuhn's own intentions in making the comparison. Consequently, though scientific revolutions might be like political revolutions in a number of important respects, this does not mean that scientific revolutions are either irrational or resolved by means of the exercise of power.

I will begin with a brief presentation of the account of scientific revolutions Kuhn develops in *Structure*. Then I present three criticisms of Kuhn's distinction between normal and revolutionary scientific changes. These criticisms draw attention to common concerns with, misunderstandings of, and ambiguities in Kuhn's view. I then examine Kuhn's mature account of scientific revolutions. In the process, I clarify what sorts of changes in science are aptly described as revolutionary. Finally, I address the concerns of the critics, and thus defend Kuhn's mature view of revolutionary changes.

THE REVOLUTIONS OF *STRUCTURE*

Despite the fact that Kuhn's view of scientific revolutions developed over the course of his life, he never revised the list of events that he regarded as scientific revolutions. The revolutionary scientific discoveries Kuhn lists in *Structure* are shown in Table 1. Included in the table are the names of the scientists who are attributed with making the revolutionary discovery, the year of their birth, as well as the year in which they publicly presented the discovery, and an indication of the book, article, or presentation in which the discovery was first reported.

These scientific discoveries differ from each other in many respects, yet when Kuhn wrote *Structure* he regarded each of them as a revolutionary discovery. That is, he regarded each as the sort of discovery that interrupts and ultimately undermines a normal scientific research tradition.

In *Structure*, Kuhn defined a scientific revolution as a "non-cumulative developmental [episode] in which an older *paradigm* is replaced in whole or in part by an incompatible new one" (1962a/1996, 92; emphasis added). Such changes are contrasted with normal scientific discoveries. Scientists making these discoveries take the accepted paradigm for granted. It is worth emphasizing that discoveries made during periods of normal science *can* be significant, for "normal" does not imply insignificant. The discovery of what has come to be known as Boyle's Law is a typical, normal scientific discovery. This discovery was certainly *significant*, but that does not make it revolutionary. There was no need to introduce a new paradigm in order to accommodate or make the discovery. Further, the evaluation of Boyle's great discovery was unequivocal. In cases of revolutionary science, evaluations are not so straightforward. This point will be illustrated in the next chapter.

In calling particular episodes in the history of science "revolutions" Kuhn sought to draw attention to similarities between a type of change in science and a type of political change, political revolutions. In *Structure*, he identifies two key similarities that warrant the comparison. First, he notes that preceding both a political revolution and a scientific revolution is a "growing sense ... that existing institutions have ceased adequately to meet the problems posed by an environment that they have in part created" (1962a/1996, 92). According to Kuhn, this "sense of malfunction that can lead to crisis is prerequisite to revolution" (92). Hence, just as a political community will not overthrow the existing institutions until they believe that these institutions are failing them, a scientific community

Table 1. *Kuhn's scientific revolutions*

Revolutionary	Revolution	Date
Copernicus	Planets orbit the sun	1514
b. 1473	*Commentariolus*: 1st draft	
Galileo	Celestial change	1610
b. 1564	*Sidereus nuncius*	
Galileo	Independence of weight and rate of fall	1616–19
b. 1564	*De motu accelerato*	
Kepler	Elliptical orbits	1609
b. 1571	*Astronomia nova*	
Boyle	Atomic theory of chemistry	1661
b. 1627	*Sceptical Chymist*	
Newton	Theory of light and colour	1672
b. 1642	Royal Society letter on light and colours	
Newton	Newtonian dynamics	1684–86
b. 1642	*De Motu*	
Musschenbroek	Leyden jar	1746
b. 1692		
Kleist	Leyden jar	1745
b. *c.* 1700		
Franklin	Theory of electrical phenomena	1750
b. 1706	"Opinions & Conjectures ..."	
Hutton	Uniformitarianism	1785
b. 1726	*Theory of the Earth*	
Black	Fixed air is distinguishable from normal air	1756
b. 1728		
Herschel	Discovers Uranus	1781
b. 1738		
Lavoisier	Discovers oxygen	1777
b. 1743		
Lavoisier	Chemical revolution	1786
b. 1743	"Réflexions sur la ..."	
Proust	Chemical law of fixed proportions	1794
b. 1754		
Dalton	Chemical atomic theory	1807
b. 1766	Edinburgh lectures	
Young	Wave theory of light	1802
b. 1773	"On the Theory of Light and Colours"	
Fresnel	Wave theory of light	1821
b. 1788		
Ohm	Ohm's Law	1827
b. 1789		
Darwin	Evolution of species by natural selection	1844
b. 1809		
Wallace	Evolution of species by natural selection	1855
b. 1823	"On the Law which has regulated the Introduction of New Species"	

Table 1. (*cont.*)

Revolutionary	Revolution	Date
Clausius b. 1822	Thermodynamics "Ueber die bewegende Kraft der Wärme"	1850
Lord Kelvin b. 1824	Thermodynamics	1851
Maxwell b. 1831	Electromagnetic theory of light "A Dynamical Theory of the Electromagnetic Field"	1865
Roentgen b. 1845	Discovers x-rays	1895
Planck b. 1858	Radiation law "Zur Theorie des Gesetzes der Energie Verteilung im Normalspectrum"	1900
Einstein b. 1879	Relativity theory	1905
Einstein b. 1879	Quantum mechanics Salzburg presentation	1909
Bohr b. 1885	Bohr's atom	1913
Heisenberg b. 1901	Matrix mechanics "Über quantentheoretische Umdeutung kinematischer und mechanischer Beziehungen"	1925

will not seek to replace its lexicon until they believe that the accepted lexicon is unfit to solve the problems that concern them. The reason why scientists are reluctant to replace the existing lexicon should be clear. "As in manufacture so in science – retooling is an extravagance reserved for the occasion that demands it" (Kuhn 1962a/1996, 76).[1]

Interestingly, a number of social scientists attribute a significant causal role to the sense of disappointment that precedes *political* revolutions (see, for example, Davies 1962; Gurr 1970; and Goldstone 1991). Perez Zagorin (1973) notes that this emphasis on the role of rising disappointments as the cause of political revolutions can be traced back to Alexis de Tocqueville (Zagorin 1973, 41). Significantly, Ted Gurr argues that it is neither

[1] Kuhn employs a variety of metaphors in *Structure*, including paradigm, religious conversion, political revolution, gestalt switch, this retooling metaphor, as well as various biological metaphors. Though his metaphors have led to interesting insights for those studying science, at times his mixing of metaphors makes it difficult to be clear what exactly he is saying about science.

deprivation in itself nor a specific degree of deprivation that leads to a political revolution (1970, 83). History provides us with many examples of people deprived of many things who do not revolt. The deprivation that seems to agitate people enough to revolt is a *relative* deprivation. That is, a precondition for a political revolution and political violence in general is a discrepancy between people's *expectations* and their *capabilities* for satisfying them. And, as Gurr explains, a person's point of reference in determining their own sense of deprivation "may be [their] own past condition, an abstract ideal, or the standards articulated by a leader [or] a 'reference group'" (25). Hence, in principle, one could feel a sense of deprivation even as one experiences an *objective* improvement in one's condition.

Assuming that the comparison with political revolutions is appropriate, a scientific theory may continue to lose followers even as the theory is being refined and is improving. This is because the theory may not be getting better fast enough, or though it is getting better, it is not improving as fast as a competitor theory is. Hence, contrary to what Imre Lakatos (1970/1972) claims, even a progressive research program may be abandoned. To ensure its status as the accepted or dominant theory, it is not enough for a theory to improve its problem-solving ability. It must improve its problem-solving ability better than competing theories. This comparative dimension of theory evaluation was a central part of Kuhn's view from the publication of *Structure*.

The second noteworthy similarity between political revolutions and these significant scientific changes is that neither sort of change is sanctioned by the existing institutions and norms. Kuhn claims that "political revolutions aim to change political institutions in ways that those institutions themselves prohibit" (1962a/1996, 93). Similarly, revolutionary scientific changes alter the existing institutions and norms in ways that the currently widely accepted theory, norms, and standards prohibit.

Not only do the scientists involved in a revolution not acknowledge the legitimacy of existing standards, the competing parties involved "acknowledge no supra-institutional framework for the adjudication of revolutionary difference" (Kuhn 1962a/1996, 93). Consequently, each party appeals to its own standards to justify their behavior. As a result, the competing parties involved in a revolution inevitably talk past each other. Moreover, any attempt to justify one's choice is apt to be circular (1962a/1996, 94). Both those defending the status quo and those seeking to initiate a radical change will justify their choice of theory by appealing to standards and norms their opponents do not accept. Indeed, it is this feature of scientific revolutions that has led a number of philosophers to think that

Kuhn believes that the resolution of revolutions is irrational (see especially Lakatos 1970/1972, 93; and Laudan 1984).

In this section, I want to examine three common criticisms that have been raised against Kuhn's account of scientific revolutions as presented in *Structure*. These criticisms challenge the way Kuhn draws the distinction between normal and revolutionary science.

First, some argue that the various changes in science that Kuhn regards as revolutionary did not affect scientists and scientific practice to the same degree, and consequently are not aptly described as being of one kind. Ernan McMullin (1993), for example, argues that we need to distinguish between what he calls shallow, intermediate, and deep revolutions. McMullin draws these distinctions in order to mitigate the threat posed by the very possibility of radical, cataclysmic changes in science, that is, deep revolutions.

McMullin regards the discovery of x-rays as a typical shallow revolution (1993, 59). He explains that with this discovery "no fundamental change in theory occurred. No troublesome anomalies were noted in advance [and] there was no prior crisis to signal that a revolution might be at hand" (59). Shallow revolutions, like this one, are quite circumscribed and leave much intact. They are of local interest only, affecting relatively few scientists.

More profound, McMullin argues, was the impact of "the replacement of phlogiston theory by the oxygen theory of combustion" (60). But even in this case, he claims, "the epistemic principles governing the paradigm debate" were "left more or less unchanged" (60). Thus, McMullin grants that there was an important *conceptual* change. But because the same epistemic principles are accepted before and after the change, McMullin believes that this type of change poses no threat to the rationality of science. After all, what counts as a good reason or adequate evidence does not change. McMullin regards this sort of change as an "intermediate revolution."

McMullin grants that *some* revolutions are in fact *deep*. But these revolutions, he suggests, are quite rare, and when they have occurred they occur over a long period of time. For example, as McMullin notes, the Copernican revolution "took a century and a half ... to consummate" (1993, 60). McMullin claims that even the revolutions that led to the acceptance of the theory of relativity and quantum mechanics are not deep revolutions (1993, 61). Given the significant differences between the various scientific changes that Kuhn regards as revolutions, McMullin questions whether the various events constitute one *type* of event (see also Bird 2000, 50–54).

Andersen *et al.* (2006) also claim that there are grades of revolution. They argue that revolutions differ with respect to the grades of incommensurability that the scientists involved experience (16). For example, they discuss a case of theory change that took place in ornithology in the 1830s that they regard as local. This change of theory was prompted by the need to account for a newly discovered species, the horned screamer, a species that was an anomaly, given the prevailing taxonomy.[2] This change of theory, they argue, affected only the specialists working in ornithology. Whatever tension the horned screamer may have caused for ornithologists until they developed the conceptual resources to classify it, knowledge of its existence had no unsettling effects for herpetologists and ichthyologists, that is, reptile and fish specialists.

Andersen *et al.* (2006) aim to show that some revolutions are quite localized and innocuous. Thus, the incommensurability associated with scientific revolutions need not and generally does not undermine the rationality of theory change. Rather, radical changes in science generally happen in a piecemeal fashion, where each stage in the process can be rationally defended. This piecemeal account of theory change is similar to Larry Laudan's (1984) reticulated model of theory change.[3]

Second, some critics claim that the various changes that Kuhn regards as revolutions are not different in *kind* from the various changes that occur during the phases that Kuhn describes as normal science. Instead, the various types of changes in science are more aptly construed as lying on a continuum. McMullin (1993), for example, argues that what we find in the historical record "is a spectrum of different levels of intractability, not

---

[2] The prevailing taxonomy, developed in the 1600s by the English naturalist John Ray, distinguished between water birds and land birds. The horned screamer, though, had attributes of both classes, and thus defied classification. In the 1830s Carl Sundevall developed an alternative taxonomy, one that could accommodate the horned screamer (Andersen *et al.* 2006, 74–75).
[3] It is important to note that Andersen *et al.* (2006) do believe that the concept "scientific revolution" is a useful one for understanding science and scientific change. In this respect, they differ from Laudan (1984) and others who accuse Kuhn of exaggerating the differences between revolutionary and non-revolutionary changes in science. But Andersen *et al.* do not think that scientific revolutions pose a threat to the rationality of theory change.
      Further, Chen and Barker claim that some taxonomic changes can be accommodated without violating the "no-overlap principle" (2000, S214). In such cases, the new and old taxonomies are commensurable. Like those critics of Kuhn who seek to mitigate the effects of non-rational factors on theory change, Chen and Barker aim to show that all conceptual changes in science can be rationally defended (S221).

just a sharp dichotomy between revolutions and puzzle solutions" (62–63). Further, McMullin argues that normal science is similar in important respects to the way Kuhn characterizes revolutionary science. In particular, he claims that "decision between rival theories is an everyday affair in any active part of science" (62). Hence, contrary to what Kuhn suggests, theory choice is not unique to scientific revolutions.

This concern has also been raised by others, including Karl Popper (1970/1972), Stephen Toulmin (1970/1972, 41–42), Alexander Bird (2000, 54–57), and Ernst Mayr (2004). Popper (1970/1972) rejects "Kuhn's typology of scientists" on the grounds that there are many gradations between normal scientist and extraordinary scientist (54). Popper suggests that though "there can hardly be a less revolutionary science than descriptive botany," in their efforts to solve their research problems botanists are often forced to engage in theoretical science (54). Thus, Popper believes that descriptive, experimental, and theoretical work "merge almost imperceptibly" (54).

Similarly, Toulmin (1970/1972) suggests that most of the alleged scientific revolutions that concern Kuhn are more aptly described as mere "conceptual incongruities." And given the frequency and small scale of these changes there seems to be little basis for distinguishing between normal and revolutionary scientific changes (see 44–45). Toulmin even recommends that we call some alleged scientific revolutions "microrevolutions" (47), thus proposing that scientific changes be construed as lying on a continuum.

Mayr claims that in biology "there is no clear-cut difference between revolutions and 'normal science'" (2004, 165). He grants that there have been some significant changes in biology, but insists that "even the major revolutions [that have occurred in biology] do not necessarily represent sudden, drastic paradigm shifts" (168). In fact, he claims that in biology "an earlier and subsequent paradigm may coexist for long periods … [and] they are not necessarily incommensurable" (168).[4]

Whereas the first criticism of Kuhn's distinction between normal and revolutionary science suggests that not all alleged revolutions affect science and scientists so profoundly, the second criticism suggests that

---

[4] Others have also sought to determine the extent to which Kuhn's model of change fits the biological sciences, given that Kuhn's background was in the physical sciences and his examples are drawn predominantly from physics, astronomy, and chemistry. For example, John Greene (1971) investigates whether Kuhn's theory of scientific change aptly describes the process of change in *natural history*. Greene argues that though "the Kuhnian paradigm of paradigms can be made to fit certain aspects of the development of natural history … its adequacy as a conceptual model for that development seems doubtful" (23).

revolutionary discoveries are not *categorically* different from normal scientific discoveries. Rather, revolutionary discoveries are those rare discoveries lying at one end of a continuum of discoveries, a continuum that includes even the most routine discoveries of normal science. This attempt to show that the various types of changes are not categorically different is also aimed at reducing the radical nature of alleged scientific revolutions. If revolutions are just those changes lying at the far end of a continuum on which most changes are indisputably rational, then perhaps they too are rational. Further, if there are very few genuinely revolutionary changes in science, they pose very little threat to the rationality of science.

The third criticism of Kuhn's distinction between normal and revolutionary science is that his two categories, normal science and revolutionary science, fail to provide us with the conceptual resources necessary to understand the variety of changes that occur in science. For example, Alexander Bird (2000) claims that the discovery of the structure of DNA "does not fit Kuhn's description of development – it originated in no crisis and required little or no revision of existing paradigms even though it brought into existence major new fields of research" (60). Bird argues that significant discoveries that lead to the development of new practices and fields, but that do not alter existing paradigms, fit into neither of Kuhn's categories. Hence, given the fact that Kuhn's framework does not provide us with the conceptual resources to understand the full range of changes in science, it is inadequate.

To summarize, critics have raised three challenges to the distinction Kuhn draws between normal and revolutionary science: (1) the changes that Kuhn regards as revolutionary changes are a diverse range of phenomena and thus do not belong in the same class; (2) the two types of changes, normal and revolutionary, are not *categorically* different; and, finally, (3) the two categories, normal and revolutionary, are not exhaustive.

Motivating Kuhn's critics is a concern to show that science is not influenced by irrational or non-rational factors to the extent that Kuhn implies (see Laudan 1984, 70–71). If truly revolutionary scientific changes are either *extremely rare* or can be shown to be similar in important respects to normal scientific changes, then science is shielded from the influence of non-rational factors.

KUHN'S REVISED ACCOUNT OF SCIENTIFIC REVOLUTIONS

In this section I want to examine Kuhn's mature account of scientific revolutions. To a large extent, Kuhn's mature account of scientific revolutions is merely a clarification of his earlier view, the view presented in *Structure*.

Kuhn came to realize that his earlier characterization of scientific revolutions in terms of paradigm changes led to a number of misunderstandings. In particular, a number of the early critics of *Structure* pointed out that Kuhn used the term "paradigm" in an imprecise fashion. Margaret Masterman (1970/1972), for example, argues that Kuhn employs the term "paradigm" in multiple ways in *Structure*, thus obscuring the explanatory work that the concept was meant to do (see also Shapere 1964/1980). The ambiguity associated with the term "paradigm" led to multiple interpretations and misinterpretations of Kuhn's view of scientific revolutions. Indeed, much of Kuhn's work after *Structure* is directed at correcting or addressing these misunderstandings caused by his use of the term "paradigm change."

As Kuhn refined his view in light of criticism, he developed an alternative characterization of scientific revolutions, one that makes no reference to paradigms. Ultimately, he characterizes scientific revolutions as scientific changes involving taxonomic or lexical changes. An example will illustrate what he has in mind. Whereas Ptolemaic astronomers used the term "planet" to denote wandering stars, that is, those "stars" that are not fixed stars, Copernicus used the term "planet" to denote a celestial body that orbits the sun. This is no small change. After all, whereas Ptolemaic astronomers did not consider the Earth to be a planet, Copernican astronomers did consider the Earth to be a planet.

Competing theories do not group things in the same way. They have incompatible ways of dividing objects into classes or kinds, and they have incompatible views about how the various kinds of objects relate to each other. According to Kuhn, a revolution always involves the *replacement* of one lexicon or taxonomy by another incompatible lexicon or taxonomy. Because a research community can change taxonomies only if a lexicon or taxonomy is already widely accepted, revolutions can occur only in *mature* fields, that is, fields that have experienced a period of normal science (see especially Kuhn 1962a/1996, chapter VII).

Not all taxonomic or lexical changes are the same. Revolutionary changes, Kuhn claims, violate the no-overlap principle. The no-overlap principle states that "no two kind terms … may overlap in their referents unless they are related as species to genus" (1991a/2000, 92). For example, the no-overlap principle implies that there are no dogs that are cats, and that all cats are animals (see Kuhn 1991a/2000, 92). As a result, Kuhn explains, "if the members of a language community encounter a dog that's also a cat … they cannot just enrich the set of category terms but must instead redesign a part of the taxonomy" (92). Kuhn believes that any development or discovery in science that requires a violation of the no-overlap principle will result in a scientific revolution. An accepted theory or taxonomy will need to be replaced by a new theory or taxonomy. On the other hand, developments in science that do not violate the no-overlap principle can be made within a normal scientific research tradition.

Despite revising the definition of "scientific revolution," Kuhn did not change his views about which episodes in the history of science count as revolutions. This is unfortunate because, as we will see in the next section, a clearly articulated notion of scientific revolution cannot be reconciled with the full list of examples Kuhn cites in *Structure*. Thus, at least one of the critics' concerns is justified.

In his developed account of scientific revolutions, Kuhn retains some of the features of the account he developed in *Structure*. Most significantly, he continued to believe that scientific revolutions were precipitated by disappointment with the prevailing standards, conceptual resources, and practices, and that advocates of competing theories lack shared standards rich enough to resolve their differences. Thus, he never did retract the comparison with political revolutions.

We are now in a position to identify the necessary and sufficient conditions for a Kuhnian revolution. For a scientific revolution to occur, (1) a research community must make a taxonomic or lexical change that violates the no-overlap principle; (2) the change must undermine the shared standards of the research community; and (3) there must be widespread disappointment with existing practices. Each of these conditions is a necessary condition for a revolution. But none on its own is a sufficient condition. It is only when the three conditions occur together that a scientific revolution occurs.

After all, a taxonomic change need not violate the no-overlap principle. Moreover, some taxonomic changes can be made without undermining the shared standards of a research community. For example, the addition of a new term designating a newly discovered animal species may be accommodated by merely extending the existing biological taxonomy, by adding an additional branch to the existing taxonomy. The discovery of many new species, including, for example, the recently discovered *Batrachylodes* frog in Papua New Guinea, can often be integrated into the accepted taxonomy by merely adding a new branch to an existing taxonomic tree (see Schenkman 2010, 301). Membership in the neighboring species is unaffected by the discovery. Such an addition neither violates the no-overlap principle nor is it apt to affect the prevailing standards of evaluation. But in the case of the change of theory that occurred with the discovery of the horned screamer, a set of concepts had to be reorganized. Ornithologists had to change the relations between the concepts such that particular things that were regarded as belonging to the same class no longer belong to the same class (Andersen *et al.* 2006, 69–75). Hence, in cases such as this one evaluating the competing theories can become problematic.

Widespread disappointment with existing practices is also necessary in every scientific revolution (Kuhn 1961/1977, 208). Unless there is such disappointment, there would be no *reason* to change taxonomies. Scientists are not whimsical. Indeed, were a research community to change taxonomies without there being widespread disappointment, science would differ little from the fashion industry where mere changes of *taste* are sufficient to cause radical change.

It is worth stressing that these conditions are conditions that affect a research community, not individual scientists. An individual's decision to adopt a theory different from the one she accepted in the past does not qualify as a revolution. Hence, when Tycho Brahe accepted a new theory of planetary motion that embodied a lexicon or taxonomy at odds with the then widely accepted Ptolemaic taxonomy, a revolution in astronomy did not occur. The relevant locus of taxonomic or lexical change is the research community. A revolution occurs only when a *research community* changes taxonomies or lexicons in the manner specified above.

This point about the locus of change in science is easy to overlook. In fact, Kuhn admits that in his earlier discussions of scientific change he often confused the attributes and experiences of individual scientists with the attributes and experiences of the research community. For example, he occasionally carelessly implies that both individuals and research communities undergo a gestalt shift when they accept a new theory (see Kuhn 1989/2000, 88). Clearly, as Kuhn recognizes, only an individual could have such an experience. A research community is not a perceiving agent capable of gestalt shifts. In attributing revolutionary changes to research communities we are not required to treat them as capable of perception or cognition.[5]

It is also worth noting that Kuhn uses the term "revolution" as a term of success only. Given his definition, there is no such thing as a failed

---

[5] When an individual scientist or a historian of science moves between an older theory and its successor, she may experience something like a gestalt shift. But a gestalt shift is not a revolution (see Barker *et al.* 2003, 220; and Nersessian 2003, 185).

scientific revolution. Though the expression or elaboration of an alternative theory may seem to indicate the potential for a revolution, and its
subsequent suppression may seem like an apt candidate for a *failed* scientific revolution, the sorts of events that Kuhn would call revolutions
require the overthrow of one theory by another. A revolution occurs only
when one lexicon or taxonomy is replaced by another incompatible lexicon or taxonomy.

Incidentally, confusions about the meaning of "revolution" are not
unique to Kuhn or philosophers and historians of science. There is still
no consensus among social scientists on the nature of *political revolutions*.
In the early 1970s, Zagorin examined a variety of definitions of "political revolution" in various theories of political revolution developed in
the social sciences and history. He found them all wanting in one way
or another, and concluded that "after this review of theories of revolution, the main conclusion to be drawn is that the subject is in a lively
but disorderly state" (1973, 52). Isaac Kraminick (1972) also examined the
competing definitions and explanations of revolution in the then recent
scholarship. He reached a similar conclusion. Kraminick argued that
"as diverse as is the literature defining revolution, there is an even wider
assortment of explanations" for why revolutions occur (35). He distinguishes between four broad types of explanations: political, economic,
sociological, and psychological. The psychological, he suggests, are the
least plausible, and all four types of explanations seem to rightly presup-
pose a political dimension.[6]

An examination of some of the most influential work on political
revolution published since the publication of Zagorin's and Kraminick's
papers reveals that little has changed. The term "revolution" continues to
be used in a variety of ways (compare, for example, Gurr 1970 with Tilly
et al. 1975; Skocpol 1979; Goldstone 1991; or Goldstone 2003). Some treat
political revolutions as a sub-class of political violence, others as a sub-
class of social revolutions.

In an effort to advance our understanding of political revolutions,
Zagorin urges social scientists and historians to reserve "the term ['revolution'] for a single, reasonably well marked out class of events" (1973,
27). Philosophers and historians of science need to show similar restraint
if the concept of "scientific revolution" is to aid us in understanding the
dynamics of scientific change. Indeed, if we are to benefit from Kuhn's

---

[6] Kraminick also discusses *Kuhn's* theory of revolution in an effort to determine what insight it
  might provide into the nature of *political* revolutions.

analysis of science it will be imperative that we use this term in the precise way he came to define it.

KUHNIAN REPLIES TO THE CRITICS

In this section I want to defend Kuhn's *revised* account of scientific revolutions from the three criticisms presented above in the section on 'Three criticisms of Kuhn's distinction'. In addressing these concerns I will be clarifying Kuhn's view, as well as making some concessions to the critics. The result is a more precise and defensible account of scientific revolutions than the one that is articulated in Kuhn's writings.

First, it must be conceded to the critics that the sorts of changes that Kuhn identifies as revolutions are a mixed lot. Some of Kuhn's examples of revolutions do not have the requisite necessary features. For example, the discovery of x-rays is not aptly characterized as a scientific revolution. Even though this discovery opened up new and unanticipated areas of research, it did not lead to the replacement of one taxonomy or lexicon by another incommensurable taxonomy or lexicon (see Chen and Barker 2000). Rather, the discovery was accommodated by creating a new field, a field devoted to the study of the hitherto unnoticed phenomena. And even though the discovery of x-rays had important implications for neighboring fields, it did not require the replacement of the taxonomies or lexicons employed in neighboring fields. X-rays could be added to the inventory of possible entities by merely adding to or extending the accepted taxonomy. Since the discovery of x-rays did not lead to the development of a new taxonomy incompatible with the old taxonomy, it is a mistake to count it among the class of revolutions.

The discovery of x-rays is not the only example of an alleged revolutionary discovery identified by Kuhn that really does not warrant the name. The discovery of the planet Uranus merely required the extension of an existing taxonomy. There was no need to replace the existing taxonomy with a new *incompatible* one. Accommodating the discovery of the new planet did not require rearranging the relations between previously known celestial objects. Hence, the critics' first concern is legitimate. The sorts of changes that Kuhn has grouped together under the label "revolution" are not all the same type of change. The scientific changes that deserve to be called revolutions always involve the replacement of one lexicon or taxonomy by another.[7]

---

[7] Some sociologists also expressed confusion about which episodes in the history of science were to count as scientific revolutions. For example, criticizing Kuhn's earlier paradigm-related account

Let us now consider the second criticism, that revolutionary scientific changes and normal scientific changes are not categorically different. McMullin, Toulmin, Mayr, and others are correct to suggest that there are *some* similarities between revolutionary and normal scientific discoveries. And some of the differences between revolutionary discoveries and normal scientific discoveries are merely differences of *degree*. For example, the sorts of problems or anomalies that lead to many sorts of discoveries, both normal and revolutionary discoveries, may appear intractable at first. What distinguishes the problems that give rise to revolutionary discoveries from the problems that one encounters during periods of normal science is that the former are generally *more* intractable. Further, no matter how fundamental a change is in science there is always some degree of consensus among the relevant practitioners. Even in revolutionary changes there is some degree of continuity with the science that preceded the change. What distinguishes periods of revolutionary science from periods of normal science is the extent of consensus in the research community. In these respects, the differences between revolutionary and normal science are differences in degree.

But in suggesting that the differences between revolutionary changes and normal changes in science are *merely* differences of degree, McMullin and the other critics imply that there is no *principled* way to distinguish the two classes of events. On this point, the critics are mistaken. In normal science, as we saw above, scientists agree about the standards by which a contribution is to be evaluated, whereas in revolutionary science the parties involved do not agree about the standards by which their competing claims should be judged. Consequently, as argued above, revolutionary changes in science are resolved in a manner that resembles the resolution of political revolutions. This point will become even clearer in the next chapter when I examine in detail the revolutionary change that took place in early modern astronomy. We will see how the disputes that lead to scientific revolutions really do give rise to distinct sets of standards or at least sets of standards that diverge and make evaluating the competing theories challenging.

---

of scientific revolutions, Edge and Mulkay (1976) argue that "it is far from easy to state unambiguously what kinds of changes should, from Kuhn's perspective, entail the occurrence of a revolution" (392). Given Kuhn's developed account of scientific revolutions, this is no longer a concern: all and only those changes that meet the conditions outlined above in the section on Kuhn's revised account are revolutionary.

Having established that there is a categorical difference between revolutionary changes and normal changes in science, and granting that some of the episodes Kuhn discusses as revolutions are not in fact revolutions, we can understand why Kuhn insists that McMullin is mistaken in distinguishing between deep and shallow revolutions (see Kuhn 1993/2000, 251). Kuhn explains that "though revolutions differ in size and difficulty, the *epistemic problems* they present are … identical" (251; emphasis added). They really do form a set worth distinguishing from other related phenomena. And they share a set of epistemic problems.

Let us now consider the critics' third concern: normal science and revolutionary science do not account for all of the types of changes in science. Bird is certainly correct about this. But the Kuhnian account of scientific change provides us with greater resources to account for the range of changes in science than Bird claims. Kuhn's account of the developmental cycle of scientific change explicitly recognizes at least two additional types of changes, what, in the language of *Structure*, would be called paradigm-creating changes and pre-paradigm discoveries. Strictly speaking, these should be called "theory-creating" discoveries and "pre-theoretic" discoveries, for in his later writing Kuhn distinguishes between theories and paradigms.

In chapter 3, we will see that the London conference in the 1960s played a crucial role in causing Kuhn to rethink the relationship between paradigms and theories. At that conference Masterman (1970/1972) insisted that sometimes in the stage that Kuhn had initially described as pre-paradigm, that is, before there is a widely accepted *theory*, those working in a research area often have a *paradigm*, that is, a concrete exemplar, to guide them in research. In fact, Masterman suggests that sometimes the various competing schools each have their own paradigm or exemplar, even though they may lack a theory (see 73–74). In Kuhn's response to his critics at the London conference, he acknowledges that Masterman is correct about this (see Kuhn 1970b/2000, 167–68). And after the London conference Kuhn came to realize that it is inappropriate to refer to the phase of research in a nascent field as the pre-paradigm phase. The early researchers in a field often have paradigms or exemplars to guide them in research. What they lack is an agreed-upon theory or lexicon. They do not yet divide the things in the world the same way (see Kuhn 1970b/2000, 168–69n. 58). The need to change the name of that phase of scientific research became even more pressing when Kuhn came to characterize theory changes as involving taxonomic or lexical changes, rather than paradigm changes.

Kuhn thus recognizes a class of scientific discoveries that occur during the pre-theory stage of a field, that is, the time before a field has developed its first *theory*. The sorts of discoveries made at this time are instances of neither revolutionary science nor normal science. Such discoveries neither involve the application of concepts from a widely accepted taxonomy or lexicon, as in normal science, nor aim to replace a widely accepted taxonomy or lexicon, as in revolutionary science. Kuhn discusses this class of scientific discoveries in his analysis of the various discoveries made by those who studied electrical phenomena before a research community was formed, that is, before a single lexicon was widely accepted (see Kuhn 1962a/1996, 13–14). Different groups of scientists interested in electrical phenomena worked with different sets of concepts. There was no scientific lexicon accepted by all those studying electrical phenomena. During this phase of the nascent field a wide range of research practices, instruments, and standards was employed. Still, some discoveries were made despite the lack of a common lexicon accepted by all those studying electrical phenomena. It is the accumulation of a number of apparently disparate discoveries that ultimately leads to the creation of a unifying theory, one that can find order in the various phenomena, or at least a substantial sub-set of them.

Kuhn also recognizes a class of changes that are aptly called theory-creating changes. A discovery that leads to the creation of the first theory in a field is also neither a revolutionary change nor a normal change. Bird's example of the discovery of DNA seems to fit this description. The discovery of x-rays also fits this description. Unlike the taxonomic change that occurred when Descartes' mechanistic physical theory *replaced* the late Renaissance Aristotelian physical theory, the discoveries of x-rays and DNA did not involve the replacement of one theory by another. Rather, in these two cases a new theory was created that led to the creation of a new scientific specialty. Hence, the critics are mistaken. Kuhn provides us with ample resources to account for the variety of discoveries made in science. I will discuss specialty-creating discoveries, like the discovery of x-rays, in greater depth in chapter 7. Kuhn's remarks in the last fifteen years of his life on this particular class of changes in science are quite insightful and significantly advance our understanding of both specialization in science and scientific change.

In summary, I have shown that Kuhn provides us with a principled way to distinguish revolutionary from non-revolutionary scientific changes. Scientific revolutions are those changes in science, precipitated by disappointment with existing practices, involving the replacement of one lexicon by another, that cannot be resolved by appealing to shared standards. I have argued that these are an important class of changes in science. They have certain features that make them especially interesting to philosophers concerned with the epistemology of science. In particular, the choice between competing theories must be made without the aid of shared standards, something that is taken for granted during periods of normal science.

CHAPTER 2

The Copernican revolution revisited

Are there *any* scientific revolutions? Such a question has been raised by some of Kuhn's critics. Larry Laudan (1984), for example, argues that *no changes* in science are aptly described as *revolutionary*. Laudan claims that given Kuhn's holistic account of paradigms a revolutionary change involves simultaneous changes in methods, goals, and theories. But, according to Laudan, Kuhn's holism "leads to expectations that are confounded by the historical record" (84). Laudan argues that all changes in science are continuous enough with the traditions preceding them to make calling any of them "revolutions" inappropriate. The only reason one would be led to believe otherwise, he claims, is if one fails to look at the process of change in sufficient detail. This same criticism, which challenges the very existence of revolutions in science, was also raised earlier by Stephen Toulmin. Writing in the 1960s, Toulmin claims that "students of political history have now outgrown any naïve reliance on the idea of 'revolutions'" (1970/1972, 47). Similarly, he argues, "the idea of 'scientific revolution' will have to follow that of 'political revolutions' out of the category of explanatory concepts" (47).[1]

My aim in this chapter is to address this criticism. I aim to show that there really are Kuhnian revolutions in science. In an effort to argue my case, I will show how the concept of a Kuhnian revolution provides insight into the change that occurred in early modern astronomy, a change that has come to be called "the Copernican revolution." In addition to showing that there are in fact scientific revolutions, I will demonstrate the value of the concept "scientific revolution," and the aptness of Kuhn's comparison of theory change to radical political changes. This case study is also important because this particular historical episode was important to Kuhn's

---

[1] It is worth noting that Toulmin does not cite a single source from contemporary political history suggesting that the concept "revolution" has become obsolete or dispensable. Though there is still much debate about the nature of political revolutions, contrary to what Toulmin claims, the concept continues to play a key role in history and the social sciences.

own thinking about scientific change. Five years before the publication of *Structure* Kuhn published a book-length treatment of the Copernican revolution in astronomy. This book was published without the benefit of the detailed account of theory change Kuhn presents in *Structure*. Consequently, Kuhn's pre-*Structure* study of the Copernican revolution is rather unKuhnian in some respects.[2] Further, given that Kuhn revised his understanding of scientific revolutions, it is important to determine the extent to which this episode fits Kuhn's developed account of revolutionary change, the account articulated in the previous chapter.

CHALLENGES TO THE COPERNICAN REVOLUTION

A number of historians and philosophers of science question whether there was in fact a Copernican revolution in astronomy. Some have raised questions about what sorts of changes the revolution in astronomy involved. Though some suggest it involved a change in theory, others suggest that it involved a change in practice. And others focus on the introduction of new instruments. Further, some have questioned whether Copernicus is rightly regarded as the astronomer responsible for the revolutionary change, whatever the change is that is rightly identified as the revolution in early modern astronomy.

I. B. Cohen (1985), for example, suggests that if there was a revolution in sixteenth-century astronomy it involved the changes in *practice* introduced by Tycho Brahe. Brahe, after all, introduced a number of important practices into observational astronomy. He collected observations over many nights, averaging over eighty nights of observation each year. Brahe also employed teams of observers who could check each others' results. And he had enormous and precise instruments made in order to ensure that he was able to achieve a degree of accuracy hitherto unattained. As far as Cohen is concerned, the sixteenth century witnessed no change in *theory*. And, compared to Brahe, Copernicus looks like an amateur. In fact, Copernicus included only about thirty of his own observations in his great contribution to astronomy.

Olaf Pedersen (1980) holds a similar view, noting in addition that Copernicus "adhered more strictly than Ptolemy to the principle of uniform, circular motion as the only admissible mathematical tool for the

---

[2] Peter Barker (2001b) argues that Kuhn's treatment of the Copernican revolution "in *The Structure of Scientific Revolutions* … is uneven and not well integrated with the main theses of the book" (241).

theoretical astronomer" (694). Indeed, Copernicus' research was motivated, in part, by the desire to construct models of the planets' motions
that did not employ equant circles, devices that were introduced by
Ptolemy that clearly violated the principle of uniform circular motion.
Similarly, Michael Heidelberger (1976/1980) suggests that "*with the emergence of Copernicus' theory, no paradigm-shift occurs but rather a coalescing
of two traditional paradigms*" (277). Hence, as far as Heidelberger is concerned, the change that occurred is not aptly described as an instance of
theory replacement. Rather, he describes Copernicus as "a faithful adher-
ent to the tradition," that is, the *Ptolemaic tradition* (279). Peter Barker
(2001b) holds a similar view, seeing more similarities between Copernicus
and Ptolemy than between Copernicus and Kepler (260, 269, 271–72n.
13). Similarly, Andersen *et al.* argue that "Copernicus' work can be seen
as a minor variation on the conceptual structure in astronomy established
by Claudius Ptolemy" (2006, 4–5).

Angus Armitage (1957/2004) also suggests that "it is possible to regard
Copernicus as the last of the ancient astronomers rather than the first
of the modern ones" (176). Kepler, he argues, is the true revolutionary.
Andersen *et al.* (2006) also claim that Kepler was the real revolutionary, for he introduced a new type of concept into astronomy, an "event"
concept rather than a "object" concept (see also Barker 2002 and 2001b).
Indeed, Barker argues that a number of the innovations in early modern astronomy that are commonly attributed to Copernicus are actually
Kepler's contributions (see Barker 2002, 208–09).

Some who have regarded Kepler as the *real* revolutionary in early modern astronomy appeal to the fact that Copernicus was a timid sort of person, as evidenced by his reluctance to publish. He was thus not fit to be a
revolutionary scientist. This interpretation, based on Copernicus' character or psyche, is central to Arthur Koestler's (1959) account of the history
of early modern astronomy, and has been uncritically repeated by others
in at least one popular textbook (see, for example, McClellan and Dorn
1999, 208–14).[3]

When Kuhn wrote *The Copernican Revolution* he was aware of some
of these criticisms and concerns. Noting that "Copernicus is frequently
called the first modern astronomer," Kuhn acknowledges that "an equally

---

[3] Robert Westman (1994) raises additional concerns with Kuhn's treatment of the Copernican
revolution. He suggests that Kuhn has a presentist "view of scientific disciplinarity" (88). This
is evident from the fact that Kuhn largely neglects astrology in his narrative and analysis (see
Westman 1994, 89). Although astrology is irrelevant to contemporary astronomy, Westman
insists that astrology was intimately linked with astronomy in the early modern period.

persuasive case might be made for calling him the last great Ptolemaic astronomer" (1957, 181). Further, like Barker, Kuhn recognized that many of the contributions that we associate with the Copernican revolution, including "easy and accurate computations of planetary position, the abolition of epicycles and eccentrics, the dissolution of spheres ... [and] the infinite expansion of the universe ... are not to be found anywhere in Copernicus' work" (135). But it is worth remembering that when Kuhn wrote *The Copernican Revolution*, he thought of scientific revolutions neither as lexical changes nor paradigm changes. Rather, he seemed then to identify the revolutionary status of Copernicus' book with the fact that it presented the next generation of astronomers with new problems (183).

RE-EXAMINING THE REVOLUTIONARY STATUS OF THE COPERNICAN REVOLUTIONARY

In this section we will see that the changes that took place in astronomy in the sixteenth and seventeenth centuries can be explained by Kuhn's developed account of theory change. A revolutionary change did occur in astronomy. It was initiated by Copernicus. And Kepler, Galileo, and others played a key role in bringing the research community around to accepting the new theory.

As stated in chapter 1, a revolutionary change involves a significant taxonomic or lexical change in a research community, where there is dissatisfaction with the accepted theory and the conflict cannot be resolved by appeal to shared standards. Hence, the locus of scientific revolutions is the research community, not the individual scientist. Some scientists, though, do play a special role in the process of theory change that warrants calling them revolutionaries.

I will argue that Copernicus should be seen as responsible for *initiating* the scientific revolution that bears his name. But Copernicus should not be seen as bringing about the revolution in early modern astronomy on his own. This is because the revolution took some time, at least seventy years. Armitage, Andersen *et al.*, and Barker are right to give some credit to Kepler. In fact, both Kepler and Galileo should be seen as playing a key role in leading the research community to accepting the new theory. Hence, individual scientists should be described as revolutionary if either (1) they are the first to propose lexical changes that violate the no-overlap principle that are ultimately accepted in their field, or (2) they play a key role in bringing the research community around to accepting the new theory. With respect to the changes in early modern astronomy, Copernicus counts as a revolutionary scientist for the first reason, whereas Galileo and Kepler count as revolutionaries for the second reason.

The changes that took place in astronomy in the sixteenth and seventeenth centuries satisfy all three of the criteria for revolutionary change as identified above. Copernicus' theory of planetary motion involved a significant taxonomic or lexical change, a hallmark of a scientific revolution. In particular, the meaning of key terms in astronomy was altered, most notably "planet." To repeat what was noted earlier, whereas Ptolemaic astronomers regarded "planets" as wandering stars and did not count the Earth as a planet, Copernicans grouped the Earth with Mercury, Venus, Mars, Jupiter, and Saturn, and conceived of planets as celestial bodies that orbit the Sun. Copernicus also introduced a new *type of entity* into astronomy, satellites of planets, though he did not use the term "satellite." Rather, it was Kepler who introduced the *term* "satellite." What is important, though, is that the *concept* of a satellite, that is, a celestial body orbiting another celestial body, is an integral part of Copernicus' theory. According to Copernicus' theory the Moon orbits the Earth as the Earth orbits the Sun. In Ptolemy's theory, no celestial body orbits another body that simultaneously orbits a third body. Hence, the changes that occurred in astronomy in the sixteenth and seventeenth centuries involved significant lexical changes. And such changes could not be made by simply extending the then widely accepted Ptolemaic lexicon. This means that the change in astronomy involved more than a change in practices or a coalescing of two traditional paradigms, as Heidelberger claims.

Figures 1 and 2, showing the Ptolemaic and Copernican taxonomies of celestial bodies, illustrate the magnitude of the changes wrought by Copernicus.

Most noteworthy are the following three changes:

- Copernicus subsumed the Earth under the category of celestial bodies, in particular, under the sub-category planet. Hence, the Earth was no longer a unique body.
- Copernicus introduced a new kind of entity, a satellite, that is, a body that orbits another orbiting body. Although initially the Moon was the only such body, Galileo's telescopic observations in the first decades of the seventeenth century would reveal additional satellites, the so-called "Medicean stars" that orbit Jupiter.
- The number of planets was reduced by one. Two bodies that were previously regarded as planets, the Sun and the Moon, were reclassified as a star and a satellite, respectively. And one body, the Earth, which was not regarded as a planet, was now regarded as a planet.

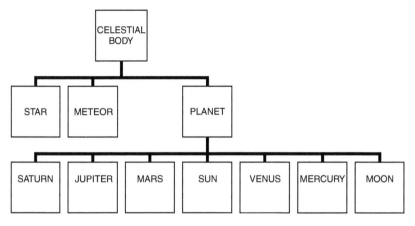

* **Note**: The Earth is *not* a celestial body according to the Ptolemaic theory.

Figure 1 The Ptolemaic taxonomy of celestial bodies*

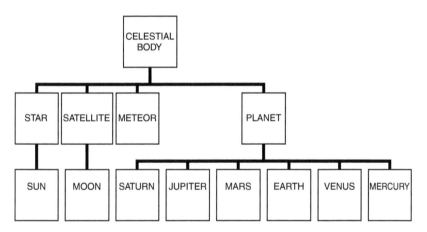

Figure 2 The Copernican taxonomy of celestial bodies

Some who have disputed the existence of Kuhnian revolutions believe that there are always shared standards rich enough to resolve disputes in science. A careful look at the changes in early modern astronomy, though, will show that the *shared* standards were not rich enough to resolve the dispute in a straightforward and unequivocal manner.

It is easy to understand why some might be led to think that early modern astronomers *did* agree on standards. No matter what their theoretical

allegiances were, early modern astronomers agreed that an acceptable theory should allow us to make accurate predictions of the locations of stars
and planets, as well as the dates of eclipses. Because accuracy in predictions can be quantified, it may seem that early modern astronomers did
in fact accept a common standard.

If the dispute in astronomy *could* have been resolved by appeal to
quantitative measures alone, then the comparison with political revolutions would be inappropriate. But as a matter of fact, the quantitative
standards were insufficient for resolving the dispute. This is because the
competing theories were equally strong from a quantitative point of view.
The two theories were equally prone to error with respect to predicting
the locations of the Moon and the planets, and the margins of error of
the two theories were also comparable. Owen Gingerich explains that
"in the Regiomontanus and Stoeffler ephemeredes [which were based on
Ptolemy's theory] the error in longitude for Mars is sometimes as large as
5°. However, in 1625, the Copernican errors for Mars reached nearly 5°"
(1975, 86). And the ranges of errors for both theories were even greater
with respect to their abilities to predict the location of Mercury (Gingerich
1971). Further, according to Victor Thoren (1967), "the Ptolemaic and
Copernican theories frequently differed by over ½° in predicting the longitude of the moon, and it was common knowledge that the moon was
rarely to be found in the place assigned to it by either theory" (21).[4] Thus,
the shared quantitative standards underdetermined theory choice (see
also Heidelberger 1976/1980, 274).

Even though the theories were roughly comparable from a quantitative
point of view, the reaction of astronomers to Copernicus' work varied.
Some astronomers wholly rejected his theory. Others, like the Wittenberg
school, were somewhat ambivalent. According to Robert Westman (1975)
and Barker (2001b), the astronomers at the university in Wittenberg were
quick to employ Copernicus' *methods* even though they rejected his innovation in *cosmology* (see Barker 2001b, 260). Westman (1975) claims that
"the principal tenet of the Wittenberg interpretation was that the new
theory could only be trusted within the domain where it made predictions
about the angular position of a planet" (166). Further, Westman notes
that for the Wittenberg astronomers "the least satisfactory Copernican
claim was the assertion that the earth moved and that it moved with more

---

[4] Gingerich (1973) provides graphs comparing the accuracy of both the Copernican and the
Ptolemaic theories with respect to their predictions for the locations of Mercury, Venus, the Sun,
Mars, Jupiter, and Saturn over a number of years (54).

than one motion" (167). But, significantly, not all astronomers adopted the Wittenberg interpretation or suspended their judgment to await the arrival of further quantitative data. Some saw promise in the new theory.

The fact that some astronomers were persuaded to accept the Copernican theory, while others remained loyal to the Ptolemaic theory, strongly suggests that *qualitative* considerations had a significant impact in resolving the dispute. Qualitatively, the theories differed significantly. Each theory could address different problems, and neither theory addressed all problems. Copernicans could explain why Venus and Mercury do not depart far from the Sun, and later, when the telescope was employed in astronomy, why Venus exhibits the same range of phases as the Moon. Heidelberger (1976/1980) notes a number of other significant differences between the Ptolemaic and Copernican theories that speak in favor of the Copernican. With the Copernican theory,

one can ... explain why the retrograde motion appears greater, the closer the planet is to the earth, and why this is true for the outer planets only when they are in opposition and for the inner planets only when they are in lower conjunction ... [and] why exterior planets seem brightest in opposition. (275)

Heidelberger also notes that the Copernican theory can "account for the peculiar way in which the sun governs the planets in the Ptolemaic system" (1976/1980, 275; see also Hoskin 1997, 47). All of these phenomena are to be expected, if, as Copernicus claimed, the Earth and the other planets orbit the Sun. These qualitative differences would incline some astronomers to accept the Copernican theory even though it was no more accurate than the Ptolemaic theory, and even though it seemed to conflict with the accepted Aristotelian terrestrial physics.

Ptolemaic astronomers, on the other hand, had a theory that fit better with a literal interpretation of the Scriptures. Moreover, the Ptolemaic theory fit better with the then accepted terrestrial physics, a version of Aristotle's physical theory. Ptolemaic astronomers were also able to cite the fact that stellar parallax could not be detected, a phenomenon that seemed to be implied by Copernicus' theory, but not by their own theory.

Early modern astronomers disagreed about the standards by which a theory should be judged. This is a consequence of the fact that they did not agree about what problems an acceptable theory *should* address. Early advocates of the Copernican theory regarded the strengths of the Copernican theory as more significant than the strengths of the Ptolemaic theory. In contrast, those who remained loyal to the

Ptolemaic theory regarded the strengths of that theory as more significant than the strengths of the Copernican theory. Further, there was
no agreed way to order and weigh the considerations in favor of each
theory (see Kuhn 1977c, 322–25). Hence, the dispute in early modern
astronomy was resolved, to a large extent, without the aid of shared
standards.

Given the protracted nature of the lexical change that occurred in early
modern astronomy, some critics find it odd to call the event a "revolution." As we saw earlier, McMullin suggests that the Copernican revolution took 150 years to run its course. Betty Jo Teeter Dobbs (2000) also
expresses concern about the fact that many alleged scientific revolutions,
the Copernican revolution and *the* Scientific Revolution in particular, are
such long-drawn-out affairs (31). Clearly, she claims, events extending
over such a long period of time hardly deserve to be called revolutions.
Cohen (1985) also raises this concern. Cohen regards revolutions as essentially abrupt and significant changes.

But there is no reason to believe that revolutions need be abrupt
events. Richard Westfall (2000) claims that the key similarity between
significant scientific changes and political revolutions that warrants
treating the former as "revolutions" is the thoroughness of the changes,
rather than the brevity of the events (44). Kuhn would certainly agree.
Further, it is worth noting that even political revolutions are not necessarily abrupt. Perez Zagorin, for example, claims that "in 1850, Alexis
de Tocqueville [noted that] ... 'for sixty years we have been deceiving
ourselves by imagining that we saw the end of the [French] Revolution'"
(Zagorin 1973, 24). Given Kuhn's developed account of scientific revolutions, the length of time that it takes for a research community to
change its theory is not a salient feature. More relevant is the nature of
the change, and the way it is resolved. Even in *Structure* Kuhn explicitly
claimed that revolutions can take a generation to run their course (see
Kuhn 1962a/1996, 166).

Critics who emphasize the fact that the change in theory in early modern astronomy was drawn out seem to assume that if the *individuals*
involved experience the change as continuous with the preceding tradition, then there is little basis for calling the event a revolution. Indeed,
Laudan's (1984) reticulated or gradualist model of scientific change is
intended to show how alleged revolutions in science are actually constituted by a series of non-revolutionary changes. Kuhn's mistake, Laudan
suggests, is due to the fact that he fails to examine the events in enough
fine-grained detail.

This line of reasoning, I believe, is mistaken and betrays a fundamental misunderstanding of the nature of Kuhnian revolutions. According to Kuhn, the *locus* of scientific change is the research community. That is, it is a research community that undergoes a revolutionary change, not an individual scientist. Kuhn claims that a revolution involves "a certain sort of reconstruction of group commitments" (Kuhn 1969/1996, 181). Individual scientists can and do adopt new lexicons, but such events do not constitute scientific revolutions. Brahe's decision to abandon the Ptolemaic theory for his own new theory may have required a radical shift in the way *he* saw the world, but there was no revolution in astronomy until a new lexicon came to be widely accepted in the research community.

It should now be clear that Copernicus' modesty is also irrelevant to his status as a scientific revolutionary. Given Kuhn's definition of a revolutionary change of theory, the character of a scientist is irrelevant to the assessment of whether a view or theory is revolutionary. What matters is whether the change requires a significant revision of the accepted lexicon or taxonomy employed in the field. Incidentally, it is worth noting that there is evidence that many great scientists are quite modest, often even avoiding challenging those with whom they have competing priority claims to a discovery (see Merton 1959). In this respect, a modest demeanor is not inconsistent or at odds with being a revolutionary scientist. Hence, Copernicus' character is irrelevant to an assessment of his contribution to astronomy.

Kuhn's focus on the scientific research community not only directs attention away from individual scientists, it also directs attention away from theories, the traditional central object of concern for philosophers of science. Theories still matter for Kuhn, but the community is more fundamental. Kuhn's epistemology of science is a social epistemology. We can only hope to understand a change of theory if we examine the social changes that make possible the replacement of an old theory by a new one.

When we look at the change in early modern astronomy from the perspective of the research community, we can see the revolutionary nature of the change. The illusion of continuity is a function of scale. Revolutions are macro-level phenomena and appreciating the process requires a macro-level perspective. When we adopt the macro-level perspective on scientific change, we see that the Copernican revolution is a quintessential scientific revolution and that Kuhn's political metaphor is appropriate.

Indeed, even in *The Copernican Revolution* Kuhn recognized how challenging it can be to recognize a scientific revolution, even for those living in the midst of one. Kuhn compared Copernicus' contribution to a bend in the road. The bend looks continuous with both the part of the road that precedes it and the part of the road that follows it. But, on either side of the bend, it is clear that the road has gone in a new direction (see Kuhn 1957, 182).

WAS THERE *REALLY* A CRISIS IN EARLY
MODERN ASTRONOMY?

This brings us to a final consideration that critics cite in their efforts to show that the change in early modern astronomy was not a Kuhnian revolution. It is often noted that when Copernicus developed his theory, astronomy was *not* in a state of crisis (see Heidelberger 1976/1980, 275). If this is true, then one of the necessary conditions for a Kuhnian revolution is not met.

An examination of Copernicus' own work gives us some indication of the state of astronomy during his lifetime. In the Preface to his *De revolutionibus*, Copernicus explains that:

[I]n setting up the solar and lunar movements and those of the other five wandering stars, [mathematicians] do not employ the same principles, assumptions, or demonstrations for the revolutions and apparent movements ... Some make use of homocentric circles only, others of eccentric circles and epicycles. (1543/1995, 5)

Thus, as Copernicus correctly notes, there was no consensus among his contemporaries. Moreover, Copernicus was not alone in his assessment of the situation. Kuhn notes that Domenico da Novara, an astronomer working in the later years of the fifteenth century, "held that no system so cumbersome and inaccurate as the Ptolemaic ... could possibly be true of nature" (see Kuhn 1962a/1996, 69).

That there was dissension and significant disagreements among early modern astronomers in Copernicus' time is further supported by the following consideration. "From the time of Copernicus' education throughout the remainder of the sixteenth century," there were two competing research programs in astronomy (Barker 1999, 345). One research program was rooted in natural philosophy and privileged Aristotle over Ptolemy. The other research program was rooted in mathematical astronomy and privileged Ptolemy over Aristotle (see Barker 2002, 210). Neither research program, though, was fully satisfactory. In fact, the weakness of each was the strength of the other. The Averroist natural philosophers privileged

Aristotle. They invoked homocentric circles in their models. But they failed to develop astronomical models "that met contemporary standards of positional calculations" (Barker 1999, 345). The Ptolemaic astronomers, on the other hand, valued accuracy, and they employed epicycles in their models. But they failed to develop "a *natural philosophy* that met contemporary standards for physical reasoning about celestial motions" (Barker 1999, 345; emphasis added). The epicycles in the Ptolemaic models were widely regarded as physically impossible.

In 1543, when Copernicus published *De revolutionibus*, there was not yet a *crisis* in astronomy. But nor were astronomers fully content with the accepted theory. And once the Copernican research program gained adherents, discontent grew among astronomers. Astronomers became aware of the fact that Copernicus' theory could answer questions that the Ptolemaic theory could not. For example, as noted above, Copernicus' theory provides an explanation for why Venus and Mercury do not stray far from the Sun (see Copernicus 1543/1995, 19–21). On the Copernican model, the orbits of Venus and Mercury lie between the orbit of the Earth and the Sun. Hence, viewed from the Earth they cannot stray far from the Sun. Ptolemaic astronomers were aware of the facts. They knew the maximum angles of elongation of the orbits of Venus and Mercury. They even had a solution to the problem, but their solution to the problem was ad hoc. They merely *stipulated* that the center of the epicycles of both Venus and Mercury always remain on a straight line running from the Sun to the Earth (see Hoskin 1997, 47). Indeed, this stipulation solves the problem, ensuring that the theory agreed reasonably well with appearances, but once astronomers began to compare the Ptolemaic solution with the Copernican solution, the Ptolemaic solution struck a number of astronomers as inadequate. Thus, Copernicus' proposal sowed the seeds of discontent for a number of astronomers raised in the Ptolemaic tradition. That is, not only was Copernicus' theory born in an environment of disagreement, it also gave rise to further disagreements. Moreover, the discontent among astronomers increased over time. In the second decade of the seventeenth century, after Galileo announced his discovery that Venus exhibits phases like the Moon, even more astronomers became dissatisfied with the Ptolemaic theory.

The Ptolemaic method of accounting for the constrained orbits of Mercury and Venus was not the only ad hoc feature of the Ptolemaic theory that attracted the attention of astronomers after Copernicus published his theory. In the mathematical models of the Ptolemaic theory, the line between a superior planet – that is, Saturn, Jupiter, and Mars – and the center of its epicycle was always parallel to the line between the Earth and the Sun (see Hoskin 1997, 47). There was no *physical* explanation for why this should be so. Copernicus' theory provided a natural explanation for this coincidence, for on his theory the planets orbit the sun. Consequently, their motion, as observed from the Earth, is tied to the motion of the Sun. Again, this natural explanation provided by Copernicus' theory further eroded the confidence of some adherents of the Ptolemaic theory. Such subtleties probably escaped the attention of many early modern astronomers, but some did take notice of these considerations.

It is worth highlighting some of the similarities between the situation in early modern astronomy and the dynamics of political revolutions. Political revolutions do not occur at the first expression of an alternative incommensurable political ideal. Hence, the mere expression of an alternative view should not be identified as a political revolution. In fact, there may not even be widespread discontent when an alternative political ideal is presented. For a political revolution to occur, the alternative view needs to gain adherents. Similarly, a revolution did not occur in astronomy just because Copernicus proposed an alternative theory. His theory, though, did begin to breed discontent among astronomers which then led others to develop the new theory in ways that increased the discontent. In fact, the revolution was not complete until a *much-altered* Copernican theory replaced the Ptolemaic taxonomy. Hence, Barker (2000, 2001b) and Andersen *et al.* (2006) are correct that Kepler played a key role in what we now call the Copernican revolution in astronomy. Copernicus' radical new scientific lexicon may not have caught on, at least not as soon as it did, if Kepler had not made his significant refinements and modifications to the Copernican theory. Galileo's many contributions were also indispensable. Armed with the newly invented telescope, Galileo made a variety of discoveries that both challenged the assumptions of the Ptolemaic theory and supported Copernicus' theory. In addition, Galileo's work in physics and hydrostatics helped further erode the grip Aristotle had on the minds of seventeenth-century astronomers and scientists.

Barker (2001b) makes the important point that the discipline of astronomy itself, and its relationship to neighboring fields, like cosmology, was in flux in this period. Indeed, part of what needed to be determined was what the appropriate relationship was between these different scientific fields.

Brahe also made significant contributions to the revolution in early modern astronomy, as both Cohen (1985) and Pedersen (1980) rightly note. Most importantly, Brahe collected observations over many nights, a practice that was made possible by the fact that he employed a team of assistants (see Barker 2000, 214). Brahe even had two groups of assistants collecting observations of the same body at the same time from different locations in order to ensure that his observations were accurate and reliable (see Dreyer 1963). These new practices and the resulting data played an indispensable role in bringing the Copernican revolution in astronomy to a close. After all, they supplied the data that led to Kepler's discovery of his first two laws of planetary motion. Still, the innovative practices that Brahe introduced into astronomy do not warrant being called a scientific revolution, at least not in Kuhn's sense, for Brahe's practices did not require a new taxonomy or lexicon.

If the concept "scientific revolution" is to be useful in advancing our understanding of science and scientific change, we need to use it in a precise way. And, given the precise sense Kuhn gives to the term in his later writings, Brahe's innovations are not revolutionary. But, with that said, it is worth reminding ourselves that "non-revolutionary" does not mean insignificant. Even changes in instrumentation, techniques, and practices can be significant. Hence, we need not and ought not treat all significant changes in science as revolutionary changes.

In summary, in examining the nature of the changes that occurred in early modern astronomy I have sought to show that the changes ushered in by Copernicus and his followers constitute an instance of a Kuhnian revolution. Thus, there is reason to believe that the concept will be a useful one for illuminating the changes that occur in other episodes of theory change.

Some critics rightly note that there would not have been a revolution in early modern astronomy if it had not been for the contributions of Kepler and others. But, equally so, there would not have been a revolution without the radical new scientific lexicon developed in the sixteenth century by Copernicus. Dying so shortly after his book was published, Copernicus was in no position to bring the revolution to a close. That task was left to the likes of Galileo and Kepler, who needed to develop the theory and collect additional data before the revolution could be brought to a close and a victory claimed for the Copernicans.

CHAPTER 3

Kuhn and the discovery of paradigms

In the previous two chapters, I both explained and offered a defense of
Kuhn's developed account of theory change. According to Kuhn's developed account, revolutionary theory changes are no longer characterized as
paradigm changes. In light of Kuhn's mature account of theory change,
it is worth examining his mature view of paradigms. To do this, it will be
worth examining how he came to discover the notion of a paradigm in
the first place.

Kuhn tells two different stories about his discovery of the concept
"paradigm." In the Preface to *Structure* Kuhn claims to have discovered
the notion of a paradigm while working at the Center for Advanced
Studies in the Behavioral Sciences at Stanford in 1958/59. Interacting
with many *social scientists*, Kuhn was struck by the differences between
the natural sciences and the social sciences. In the former, there is broad
agreement about the fundamentals of the field, whereas in the latter
there is often significant disagreement about fundamentals. Kuhn claims
that "attempting to discover the source of that difference led [him] to
recognize the role in scientific research of what [he has] since called
'paradigms'" (1962a/1996, x). Paradigms, as he explains, are "the univer-
sally recognized scientific achievements that ... provide model problems
and solutions to a community of practitioners" (x). Paradigms, Kuhn
claims, are a standard feature of the natural sciences, but not of the
social sciences.

By 1974, in "Second Thoughts on Paradigms," Kuhn was telling a different story. He explains there that:

[T]he term "paradigm" ... entered *The Structure of Scientific Revolutions* because
[he] ... could not, when examining the membership of a scientific community,
retrieve enough shared rules to account for the group's unproblematic conduct
of research. Shared examples of successful practice [that is, paradigms] could ...
provide what the group lacked in rules. (1974/1977, 318)

48

By 1974, concern for understanding the differences between the natural sciences and the social sciences drops out of Kuhn's story of discovery. Instead, he claims to have invoked the notion of a paradigm to account for the consensus necessary among scientists in a specialty in order for them to pursue their research goals effectively. Kuhn seems to have settled on this latter story about the origins of the concept "paradigm" for he repeats it again in an interview in 1995 (see Kuhn 2000b, 296).

Significantly, and not surprisingly, given that Kuhn himself tells two different stories of his discovery, there is not much agreement among scholars about Kuhn's discovery of paradigms. Paul Hoyningen-Huene offers an account of Kuhn's discovery, but fails to mention the first story that Kuhn tells about his discovery (see Hoyningen-Huene 1989/1993, section 4.1, 132–40). Alexander Bird (2000) claims that "initially Kuhn employed the exemplar idea primarily in two roles, in explaining certain features of perception and in explaining scientific change" (95–96). Sharrock and Read (2002) claim Kuhn invokes the notion of a paradigm in order to contrast early and later stages of development in a scientific field. Sharrock and Read, though, also recognize that Kuhn invokes the notion of a paradigm in order to explain the differences between the natural and the social sciences (34).

The fact that Kuhn tells two different stories about the origins of the concept gives us reason to believe that he may be mistaken about his discovery. In this chapter, I aim to offer a different history of Kuhn's discovery of paradigms, one that takes account of the complexity of the discovery process. My analysis of Kuhn's discovery of paradigms draws on Kuhn's own account of the discovery process in science. According to Kuhn, scientific discoveries are complex, often convoluted events (Kuhn 1962a/1996, 52–56). Kuhn's analysis of the discovery of oxygen, for example, illustrates how difficult it is to identify the date at which oxygen was discovered. The discovery of oxygen, he suggests, was a complex process that unfolded over a period of years (see also Kuhn 1962b/1977).[1]

I aim to show that Kuhn's discovery of paradigms was a complex process of a similar sort. For the sake of clarity, I divide Kuhn's discovery into four phases: (1) pre-Kuhnian uses of the concept; (2) Kuhn's wide-ranging and imprecise use of the concept in *Structure*; (3) his settling on the concept of the paradigm as exemplar in response to early criticism;

---

[1] The view that scientific discovery is a complex process has been developed further by Augustine Brannigan (1981).

and (4) the mopping-up process that followed in light of his settling on the narrow use of the term.[2]

Kuhn is so much identified with the notion of a paradigm one might be led to believe that he was the first to use the term in the context of science. But this is not so. Daniel Cedarbaum (1983) has traced the first use of the term "paradigm" in discussions of science to the late eighteenth century (see Cedarbaum 1983, 181). A German scientist, Georg Christoph Lichtenberg, distinguished in his day, uses the term, and apparently in a way not unlike Kuhn. Cedarbaum suggests that Wittgenstein's use of the term may have been influenced by his reading of Lichtenberg's work, and, in turn, Kuhn's reading of Wittgenstein may have had an impact on Kuhn's use of the term (1983, 187). It is interesting to note that Hans Reichenbach (1938/2006) cites Lichtenberg in *Experience and Prediction*, though not his references to paradigms. We know that Kuhn read *Experience and Prediction* for it is from a reference in it that he was led to Fleck's (1935/1979) *Genesis and Development of a Scientific Fact* (see Kuhn 1979a, viii). Paul Hoyningen-Huene and Stefano Gattei also discuss some of the early uses of "paradigm," noting pre-Kuhnian uses in the work of Neurath, Schlick, and Cassirer (see Hoyningen-Huene 1989/1993, 132–33n. 7; Gattei 2008, 19n. 65). But, as we will see shortly, Kuhn's discovery of the concept of paradigm was probably a consequence of other factors as well.

Even before Kuhn latched on to the concept of paradigm, others in *his* intellectual circle were using the *term* "paradigm." For example, in 1949, the Harvard psychologists Bruner and Postman used the term in a paper titled "On the Perception of Incongruity: A Paradigm." It was in this paper that Kuhn encountered the example of the anomalous playing cards. This is the experiment where subjects are briefly shown anomalous playing cards, like a *red* king of *spades*, or a *black* ten of *hearts*, and asked to identify what they saw. The experimental subjects initially classified these anomalies according to their expectations, which are a result of their familiarity with a typical deck of cards. Thus, a red king of spades

[2] Kuhn identifies three common *characteristics* of scientific discoveries: (1) they are initiated by anomalies; (2) there is a struggle to normalize the anomalies; and (3) they lead scientists to rethink hitherto settled knowledge (see, for example, Kuhn 1962b/1977, 172–75). Although there is a temporal thread running through these events, the reader is forewarned that the four *phases* of my analysis do not map on to Kuhn's three *characteristics* of scientific discoveries.

would be identified as a red king of hearts, for example. Only as they were exposed to these anomalous playing cards for longer periods of time did they begin to detect a "problem" and see the cards for what they *really* were.

In their paper, Bruner and Postman (1949) conclude that "perceptual organization is powerfully determined by expectations built upon past commerce with the environment" (222). This is the key point Kuhn drew from their paper when he wrote *Structure*. But, importantly, Kuhn thinks this insight about the effects of expectations on perception is relevant to developing a better understanding of the process of scientific discovery. Kuhn claims that this experiment "provides a wonderfully simple and cogent schema for the process of scientific discovery. In science, as in the playing card experiment, novelty emerges only with difficulty, manifested by resistance, against a background of expectation" (1962a/1996, 64). Scientific discoveries are often difficult, drawn-out affairs that are initially met with resistance and disbelief, *even by those who are ultimately identified as the discoverers*. Scientists' past experiences shape their expectations in ways that can make the perception of novelty difficult.

Kuhn may have come to appreciate the fact that preconceptions can be an impediment to change in science from working with James B. Conant (see Conant 1950/1965, 4). This was a crucial theme explored in the Harvard Case Histories in Experimental Science created, under Conant's direction, for the General Education science course at Harvard. These were the textbooks used in Conant's course in which Kuhn was a teaching assistant. Indeed, Kuhn's own *Copernican Revolution* grew out of his contribution to that course.

Although Kuhn discusses Bruner and Postman's psychological study in some detail in *Structure*, he did not discuss their use of the term "paradigm." On careful inspection of their article, it is not surprising that he makes no reference to their use of the term. The word "paradigm" appears only in the subtitle and in the running head. Nowhere in the article do they define what a paradigm is. Apparently, they assume the reader would know what one is. Reading the introduction of their article, one gets the sense that a paradigm might be "any large-scale statement of principles," for they claim that is precisely what is missing in the literature on the effects of expectations on perception (see Bruner and Postman 1949, 206).

In the early 1980s, while reflecting on the influence of his own research on perception and cognition, Jerome Bruner claimed that Kuhn "used [the anomalous playing card experiment] as a kind of metaphoric exemplar of his idea of paradigms in science in his *Structure of Scientific Revolutions"* (1983, 85). Bruner is clearly mistaken about this. In *Structure* Kuhn draws on the anomalous playing-card experiment in his efforts to explain the *process* of discovery, in particular, the fact that discoveries in science are often rather drawn-out affairs. The discovery process is often drawn out because the paradigms and theories one accepts limit one's vision, and can even prevent one from noticing certain phenomena. Incidentally, Leo Postman and Kuhn were not only colleagues at Harvard in the early 1950s, they were also colleagues at Berkeley in the late 1950s and early 1960s. Indeed, Kuhn claims to have discussed the famous anomalous playing-card experiment with Postman (see Kuhn 1962a/1996, 64n. 13).[3] It is interesting that both were using the term "paradigm" and concerned with the effects of expectations on perception.

In the 1940s the sociologist of science Robert K. Merton was also using the term "paradigm." Like Kuhn, Merton completed his graduate education at Harvard, though Merton was working in the sociology department, and their time at Harvard did not overlap. I have traced Merton's use of the term to a book review in 1941 and to two articles published in 1945. In a review of Florian Znaniecki's *The Social Role of the Man of Knowledge*, a book devoted to the sociology of science, Merton refers to the four interacting components of social systems identified by Znaniecki as a paradigm (see Merton 1941, 112). The term, though, is neither explained nor used again in the review. Four years later in an article, "Sociological Theory," Merton uses the term three times. First, in a discussion of the role and value of conceptual clarification in sociology, he describes a particular example as serving "as a paradigm of the functional effect of conceptual clarification upon research behavior: it makes clear just what the research worker is doing when he deals with conceptualized data" (1945a, 467). In this context "paradigm" denotes a particularly clear example.

Oddly, just prior to discussing *this* example, which Merton refers to as a paradigm, he discusses another sociological study in order to illustrate how "our conceptual language tends to fix our perceptions and, derivatively, our thought and behavior" (1945a, 466). The particular study Merton discusses is Edwin Sutherland's 1940 study of the concept of crime. Merton explains that Sutherland "demonstrates an equivocation implicit in criminological theories which seek to account for the fact that

---

[3] See Kuhn 1962a/1996, 64n. 13. Bruner (1983, chapter 6) provides the date of Postman's move from Harvard to Berkeley. Merton (1977) provides a detailed account of Kuhn's institutional affiliations from 1943 to 1975 (77–79).

there is a much higher rate of crime, 'as officially measured,' in the lower than in the upper social classes" (Merton 1945a, 466). But, as Merton explains, "once the concept of crime is clarified to refer to the violation of criminal law and is thus extended to include 'white-collar criminality' … violations which are less often reflected in official crime statistics than are lower-class violations – the presumptive high association between low social status and crime may no longer obtain" (1945a, 466). Merton's point is that our presuppositions about crime shape what we see as a crime, as well as what we fail to see as a crime. But in this context Merton makes no mention of the concept of paradigm. Though the term paradigm is not used here by Merton the idea that one's conceptual language fixes one's perception and thought sounds remarkably like Kuhn's view of the effects that paradigms have on scientists, the same sort of effects that Bruner and Postman claim our expectations have on our perception.

In this same article, Merton reconstructs the theoretical assumptions of Durkheim in a "formal fashion" in order to clarify "the paradigm of (Durkheim's) theoretical analysis" (1945a, 470). And, in a footnote, Merton describes "the paradigm of 'proof through prediction' as logically fallacious" (471n. 24). Given the variety of ways Merton uses "paradigm" and the lack of a precise definition, it is far from clear what Merton thought a paradigm was.

Merton also uses the term "paradigm" in another paper published in 1945, "Sociology of Knowledge" (see Merton 1945b). This paper is included in his 1973 collection *The Sociology of Science: Theoretical and Empirical Investigations*, but it is given a new title, "Paradigm for the Sociology of Knowledge" (see Merton 1973). In this paper, he lists five questions for sociologists interested in studying knowledge from a sociological point of view. He describes the list as a "paradigm," a "scheme of analysis" which is intended to provide "a basis of comparability among the welter of studies which have appeared in [the sociology of knowledge]" (1945b, 371). This paradigm, he claims, "serves to organize the distinctive approaches and conclusions in this field" (373).

Later, Merton would use the term "to refer to exemplars of codified basic and often tacit assumptions, problem sets, key concepts, logic or procedure, and selectively accumulated knowledge that guide inquiry in all scientific fields" (Merton 2004, 267). For example, he would use the expressions "paradigm for functional analysis in sociology" and "paradigm for structural analysis in sociology" (see Merton 1996). Much like Kuhn's various uses of the term in *Structure*, Merton's uses of the term lack precision. At the end of his life, while reflecting on his own use of the term, Merton notes that "despite the manifest overlap with the concept of paradigm as it emerged in the 1962 *Structure* ... it is quite evident that Tom Kuhn had no idea of my usage of the term" (Merton 2004, 267).

Kuhn recalls that he first encountered Merton's writings in 1947, when he began working as an assistant for Conant in the General Education science course at Harvard. But it was Merton's thesis about the effects of Puritan values on the development of modern science that Kuhn read then, not Merton's sociological work, where he employed the term "paradigm" (see Kuhn 2000b, 287–89). Incidentally, Merton reports that while he was still a graduate student at Harvard in the 1930s, he was invited to a lunch with Conant at which they discussed the history of science (see Merton 1977, 85–86). Apparently, Conant had read Merton's paper "Puritanism, Pietism, and Science," and later read his dissertation.

In summary, the *term* "paradigm" was already in use before Kuhn began using it. Moreover, some of the people who were using the term were concerned with the same sorts of effects that Kuhn would attribute to paradigms. Specifically, both Bruner and Postman and Merton were concerned with how *prior expectations shape perception*.

Incidentally, Kuhn used the term "paradigm" even before his epiphany at the Center for Advanced Studies, in fact twice. First, he used it in his 1957 book *The Copernican Revolution*. But his use of the term in that book carries none the connotations the word would ultimately acquire in *Structure*. In the earlier book, Kuhn claims that:

> [S]ince students in this General Education course [from which the book evolved] do not intend to continue the study of science, the technical facts and theories that they learn function principally as paradigms rather than as intrinsically useful bits of information. (1957, ix)

The term "paradigm" as used here can mean nothing more than a clear example.

Kuhn also uses the term "paradigm" in his paper "The Essential Tension" (1959/1977). In this paper he seems to use the term in a fairly precise manner to mean exemplar. For example, he claims that science textbooks "exhibit concrete problem solutions that the profession has come to accept as paradigms, and they then ask the student ... to solve for himself problems very closely related in both method and substance to those through which the textbook ... has led him" (229). But by the time Kuhn came to write *Structure* such restraint had passed.

Let me refer to this as the first phase of Kuhn's discovery of the notion of a paradigm. At this stage in the process, the word is being used and there is increasing interest in the theory-ladenness of observation. But we still have a long way to go before we can say that Kuhn has discovered the notion of a paradigm. At this stage, Kuhn is in a similar situation to Priestley's in 1774. At that time, Priestley discovered that when heated, red precipitate of mercury produced a gaseous substance that "would support combustion" (see Kuhn 1962b/1977, 168–69). But he could hardly claim to have discovered oxygen.

PHASE II: THE PARADIGMS OF *STRUCTURE* AND THEIR EARLY RECEPTION

By the time Kuhn finished writing *Structure*, he was using the term "paradigm" in a variety of ways (see Hoyningen-Huene 1989/1993, 131–32). The term was used to refer to the concrete scientific achievements that guide scientists in research. The term was also used to refer to the complex sets of theories, goals, and standards that he later came to call "disciplinary matrices." And he often used the term "paradigm" interchangeably with the term "theory" such that changes of theory were referred to as "paradigm changes." This was unfortunate, as Kuhn came to realize (see Kuhn 1970b/2000, 168; 1974/1977, 293 and 319; and 1991b/2000, 221).

Early on, both Dudley Shapere (1964/1980) and Margaret Masterman (1970/1972) raised concerns about the wide range of ways the concept "paradigm" was being used by Kuhn. Shapere argued that "the explanatory value of the notion of a paradigm is suspect: for the truth of the thesis that shared paradigms are … the common factors guiding scientific research appears to be guaranteed … by the breadth of definition of the term 'paradigm'" (Shapere 1964/1980, 29). And Masterman (1970/1972) identified twenty-one different uses to which Kuhn put the term "paradigm." She distinguished between three main notions of paradigm: (1) a sociological notion, which she described as "a set of scientific habits" (1970/1972, 66); (2) an artifact or construct notion of paradigm, similar to what Kuhn would call an exemplar; and (3) a metaphysical paradigm. Although the latter notion was the principal object of criticism in philosophical discussions of Kuhn's work, Masterman believed that it was the least important of the three notions (Masterman 1970/1972, 65). Indeed, she was bothered by the fact that many of Kuhn's critics mistakenly identified a paradigm with "a basic theory" (see Masterman 1970/1972, 61). As far as she was concerned, the sociological notion and the construct notion are the keys to developing a richer philosophical understanding of science. Interestingly, Masterman describes herself as a scientist. She worked in the computer sciences (60). Thus, her lack of interest in theory is probably a consequence of the fact that she was not a philosopher of science. While most of the others at the London conference, primarily loyal Popperians, focused on criticizing Kuhn's notion of normal science, Masterman praised Kuhn for providing the conceptual resources for explaining scientific *practice*, and, especially, the work of scientists who have *no theory* to guide them (1970, 66).[4]

Kuhn identifies this insight of Masterman's as especially important in his own understanding of paradigms. He later described hearing her paper at the conference and thinking: "She's got it right! ... A paradigm is what you use when the theory isn't there" (see Kuhn 2000b, 300; see also 1970b/2000, 167–68). Thus, it is at this point that he began to conceive of paradigms as something distinct from theories.

It is worth noting that Conant, Kuhn's mentor, anticipated Shapere's and Masterman's criticism. In a letter written to Kuhn in 1961, Conant claims that he feared Kuhn would be remembered as "the man who grabbed on to the word 'paradigm' and used it as a magic verbal wand to explain everything!" (cited in Cedarbaum 1983, 173). Incidentally, Merton (2004) reports that when he began using the term "paradigm" in sociology in the 1940s, people were perplexed by his choice of terms (see Merton 2004, 267).

We are now in the second phase of Kuhn's discovery. He has embraced the word "paradigm," consciously putting it to work in his analysis of scientific discovery. But, at this point, it is far from clear what phenomenon the term is meant to designate.

Incidentally, despite Cedarbaum's intention to clarify what Kuhn meant by the term "paradigm," he seems to fall back into a broad and sloppy use of the term. For example, in a discussion of the relations between Fleck's work and Kuhn's *Structure*, Cedarbaum claims that "Kuhn's 'scientific communities' are examples of [Fleck's] 'thought collectives,' and the term 'thought style' might often be substituted for 'paradigm' in *Structure*" (1983, 194). Cedarbaum also misunderstands Kuhn when he claims that "the essential constituents of a paradigm, for Kuhn, are an axiom system and a model (in the technical sense) for that system" (Cedarbaum 1983, 204). This captures neither Kuhn's conception of paradigms, nor his conception of scientific theories.

---

[4] Among *sociologists* of science, "normal science" is widely regarded as one of Kuhn's most important contributions to our understanding of science. Science is, after all, a tradition-bound activity (see Barnes 2003). Among *philosophers* of science, Joseph Rouse is the one who has given the greatest attention to Kuhn's remarks about the practice of science (see, for example, Rouse 2003).

Gattei is guilty of similar mistakes. He seems to think a crucial aspect of Kuhn's notion of a paradigm is the idea that paradigm changes are essentially discontinuous changes (see Gattei 2008, 19n. 65). Indeed, it is this feature, claims Gattei, that distinguishes Kuhnian paradigms from other philosophers' conceptions. We will see in the section of this chapter on Phase IV: the aftermath, below, that Kuhn came to believe that the crucial changes in science are *theory changes*, which are in some important respects distinct from *paradigm changes*.

It seems that the enthusiasm with which Kuhn used the term "paradigm" in *Structure*, an enthusiasm bordering on recklessness, consumed others as well. In a recently published article, Marx and Bornmann (2010) report statistics on the number of articles that use the term "paradigm" in their titles. Searching the *Web of Science*, they found that there were six uses in 1960, eight in 1961, and twenty-one in 1962, the year *Structure* was published. Not surprisingly, the use of the term continued to grow. In 1970 there were forty-five uses, in 1971 there were sixty-six uses, and in 1972 there were eighty uses. In 1980, 198 articles had the term in the title, and in 1990, 416 articles had the term in the title. In 2000, 998 articles used the term in the title, and in 2008, there were 1,372 articles with "paradigm" in the title. Despite the lack of clarity with which Kuhn initially employed the term, he clearly hit a nerve. The term resonated with natural scientists and social scientists, as well as scholars studying the sciences, that is, historians, sociologists, and philosophers of science.

PHASE III: THE CENTRAL INGREDIENT

As Kuhn reflected on these early responses to *Structure*, he was led to clarify his own understanding of what a paradigm was or could be. He recognized that if the concept of paradigm was to do any work in his analysis of science it had to be used in a precise manner. Three papers seem especially important in this phase of the discovery process: (1) his reply to his critics at the London conference, "Logic of Discovery or Psychology of Research?" (Kuhn 1970a/1977); (2) the Postscript to the second edition of *Structure* (1969/1996); and (3) his contribution to Frederick Suppe's edited volume *The Structure of Scientific Theories*, "Second Thoughts on Paradigms" (1974/1977).

The central ingredient of Kuhn's concept of a paradigm that emerges at this time was the idea of a concrete accomplishment that serves as a guide to further research, as a template for resolving outstanding problems in a field (see Kuhn 1974/1977, 307n. 16). As Kuhn sought to address the concerns raised by Masterman, Shapere, and others, he came to call these "exemplars" (see Kuhn 1970b/2000, 168). For the sake of clarity, in the remainder of this book, I will use the term "exemplars" when I mean paradigms in this narrow sense. Crucially, exemplars must (1) be widely accepted solutions to concrete problems, but also (2) provide guidance to scientists as they try to solve other, related problems. Hence, they are not merely clear examples. Paradigms play an indispensable role in guiding future research.[5]

One of the clearest cases of an exemplar is Kepler's mathematical model that describes the orbit of Mars presented in his *New Astronomy*. That solution provides the key to solving related problems, including modeling the orbits of other planets and the orbit of the Moon. But even as one works with Kepler's solution as a guide, solving these other problems can be challenging. One cannot just *derive* an answer to these problems from Kepler's work. In an effort to achieve or construct solutions, a variety of parameters can be altered. One must determine, for example, the eccentricity of the ellipse that best describes the orbit one wants to model. One must also assign a speed to the planet, for Kepler's model requires only that the speed vary in such a way as to sweep out equal areas in equal times. Still, Kepler's solution for the orbit of Mars provides one with a variety of constraints that makes solving these problems easier than it would be without his solution as a guide.

Kuhn gives other examples of exemplars, discussing the way they are subsequently employed in an effort to solve other related problems. Galileo's work on inclined planes, for example, drew on his previous knowledge of the properties of pendulums (see Kuhn 1970b/2000, 170). Having

---

[5] Far too much has been made of the fact that Stephen Toulmin (1961) also used the term paradigm in *Foresight and Understanding*, which was published a year before Kuhn's *Structure*. Toulmin's use of the term is as amorphous as Merton's. Toulmin is most precise in his use of the term "paradigm" when he characterizes a paradigm as a "particular explanatory conception" (52). For example, he claims that Aristotle's paradigm of motion was "a horse-and-cart" (52). Given this paradigm, Toulmin argues, in every case of motion you analyze "you should look for two factors – the external agency (the horse) keeping the body (the cart) in motion, and the resistance (the roughness of the road and the friction of the cart) tending to bring the motion to a stop" (52). Clearly, there are similarities between this notion of a paradigm and Kuhn's exemplars. But there are also significant differences. Most importantly, Kuhn's exemplars specify *in detail*, and in concrete ways, the sorts of parameters that are not up for negotiation. Indeed, unlike Toulmin, it seems that for Kuhn the salient features of an exemplar cannot be merely qualitative. Perhaps the most fundamental difference between Toulmin's view and Kuhn's view is that the former believes that there are no genuinely revolutionary changes in science (see Toulmin 1970/1972, 47).

Kuhn claims to have deliberately avoided reading both Toulmin's *Foresight and Understanding* and Michael Polanyi's *Personal Knowledge* while he was still writing *Structure* (see Kuhn 2000b, 296–97). But it seems that Kuhn's memory has failed him again, for he cites Polanyi's *Personal Knowledge* in *Structure* (see Kuhn 1992a/1996, 44n. 1).

developed an understanding of the way pendulums behave, Galileo began to see bodies moving down inclined planes in a way he had not seen them before, noting hitherto unnoticed similarities between the two sets of phenomena. His understanding of pendulums thus functioned as a paradigm or exemplar for his study of bodies moving on inclined planes.

Kuhn also discusses the variety of forms in which Newton's second law, $f = ma$, is expressed as one solves related problems (see Kuhn 1970b/2000, 169). Kuhn (1974/1977, 299) lists a number of variations, including the following:

$$mg = md^2s/dt^2 \text{ for free fall, and}$$
$$mgSin\theta = - md^2s/dt^2 \text{ for the simple pendulum.}$$

The physicist-in-training must learn to work with these different expressions of the law, and, most importantly, learn to identify when to use which formulation.[6]

Paradigms as exemplars play a crucial role in Kuhn's philosophy of science, even as it developed later in his career. They are the means by which young aspiring scientists become scientists (1962a/1996, 10–11). A central part of the educational process in the natural sciences involves learning paradigms, and then applying them to new problems (see Kuhn 1970b/2000, 169–70). This is an essential aspect of the socialization process that brings one into a scientific research community (Kuhn 1962a/1996, 43; 1963, 349). Unless one can work with the accepted paradigms one will be unable to participate as a researcher. Indeed, Kuhn suggests that young aspiring scientists learn both the accepted theory and "how the world behaves" through learning to work with paradigms (see Kuhn 1970b/2000, 171; 1962a/1996, 46). Paradigms are thus the means by which aspiring scientists are socialized into their new roles as scientists.[7]

Paradigms also help explain why scientists are *usually* so successful in realizing their goals. A paradigm, like an accepted theory, focuses a

---

[6] There are many other examples of paradigms. Nancy Cartwright (1994/1996), for example, notes that "part of learning quantum mechanics is learning how to write the Hamiltonian for canonical models – for example, for systems in free motion, for a square well potential, for a linear harmonic oscillator, and so forth" (317).

[7] Hoyningen-Huene rightly notes that the role of exemplars in training scientists was central to Kuhn's thoughts from the time he began using the term "paradigm" (see Hoyningen-Huene 1989/1993, 134n. 14). Bird (2000) claims that it is from working with exemplars that scientists acquire new "learned discriminatory capacities" (75). Bird also rightly notes that exemplars operate at two levels. "Exemplars described at the individual level ... [provide] scientists with a personal intuition for similarity," which is essential for applying the exemplar to hitherto unsolved problems. "At the social level ... *the* paradigm of a certain field is an object of consensus among its practitioners" (Bird 2000, 79).

scientist's attention narrowly. This can be useful, ensuring that one attends to *only* those features of the world that matter. In this way, one is not overwhelmed by a torrent of extraneous information as one seeks to understand the phenomena that are the objects of one's study. One sees what one should see. The example of Kepler's paradigm illustrates this well. His solution to the orbit of Mars indicates which parameters are and which are not up for negotiation.

Paradigms also help us explain why changes of theory are often difficult experiences for scientists to endure. Like accepted theories, paradigms restrict a scientist's vision, which can even prevent her from seeing something that is before her eyes. Indeed, the scientist working with a long-accepted paradigm is like the subject of the psychology experiment who cannot see the black ace of hearts for what it is. It is in this way that paradigms create the conditions that make theory change difficult (Kuhn 1962a/1996, 64). If one cannot even conceive of a black ace of hearts, then one is not apt to see one even if it is right before one's eyes.

PHASE IV: THE AFTERMATH OF KUHN'S DISCOVERY

In his 1962 paper on the historical structure of scientific discovery, Kuhn notes that typically significant scientific discoveries "react back upon what has previously been known, providing a new view of some previously familiar objects" (Kuhn 1962b/1977, 175). That is, such discoveries often require us to rethink what we thought we knew before. This is part of the hermeneutical dimension of the *natural* sciences that Kuhn believed many, including Charles Taylor, fail to recognize (see Kuhn 1991b/2000, 222). We can see such a process happening in Kuhn's discovery of paradigms. As he clarified his understanding of *exemplars*, he set about the task of clearing up confusions that were a consequence of his earlier ambiguous uses of the term "paradigm." Most importantly, he no longer characterized changes in theory as paradigm changes. Kuhn thus came to distinguish theories from exemplars.[8]

---

[8] Hoyningen-Huene notes that after 1969 Kuhn "resumed talking of 'theories' and 'theory choice' where in SSR he usually substituted 'paradigm'" (1989/1993, 142–43). Bird (2000) notes a tension in Kuhn's account between exemplars and theories. Bird expresses the point in the following way: "there is a downplaying of *theory* in the account of exemplars that is in tension with the emphasis on *theory* change in Kuhn's descriptive project. Revolutions are primarily changes in theoretical beliefs" (2000, 86). Hoyningen-Huene notes that after 1982 Kuhn referred to the systems of concepts that we associate with scientific theories as lexicons (1989/1993, 159). Hoyningen-Huene also notes that Kuhn "stops using the term 'disciplinary matrix' after 1969" (1989/1993, 132).

Recall, as noted above, that after hearing Masterman's London conference paper Kuhn admits that there can be exemplars even before scientists have a theory. Thus, the two, exemplars and theories, must be distinct. Further, throughout his later writings he characterizes scientific revolutions as involving changes in the lexicon or theory that a research community employs. For example, he claims that scientific revolutions involve meaning change, that is, a "change in the way words and phrases attach to nature" (1987/2000, 29). Exemplars, on the other, are not captured in words and phrases.

Kuhn's refinement in his understanding of theory change did not make exemplars obsolete or irrelevant to our understanding of science. Kuhn continued to regard exemplars as the means by which theories are learned. The meaning of the key terms of a theory can be quite opaque to the young scientist-in-training. In order to acquire an understanding of what they mean, scientists must apply the terms, and exemplars play a key role in the process (see Hoyningen-Huene 1989/1993, 139).

Given Kuhn's refined understanding of theory change, where theories are distinguished from exemplars, it becomes easier to see both how theories and exemplars differ and how they are similar. The key similarity is that both exemplars and theories restrict a scientist's vision. As noted above, this restricted vision has positive and negative effects on scientists. It enables scientists to focus narrowly on what matters, or at least what is deemed to matter, but it also leads scientists to overlook things that may turn out to be important. And the most important difference between exemplars and theories is with respect to their flexibility. The taxonomies or lexicons that are replaced during episodes of theory change are essentially inflexible. To alter a scientific lexicon or taxonomy is to cause a change of theory. More precisely, a change of theory involves a modification of a lexicon in such a way as to violate the no-overlap principle (see Kuhn 1991a/2000, 94). For example, the list of entities covered by the term "planet" is not the same for the Ptolemaic theory as it is for the Copernican theory. Exemplars, on the other hand, are essentially flexible. Indeed, their value is a function of their flexibility. Kepler's model of Mars' orbit is only useful for understanding the orbits of other planets because the various parameters can be adjusted. A concrete scientific achievement can only function as an exemplar if it can be altered or modified in ways that enable scientists to solve *other* problems. Any scientific achievement that does not have this capacity for flexibility cannot function as an exemplar.

It is worth mentioning that not everyone has been pleased with the understanding of the concept "paradigm" that Kuhn finally settled on.

Many sociologists of science were intrigued by the earlier, more complex notion of a paradigm which "incorporated cognitive and social commitments" (see Pinch 1979, 440). Trevor Pinch, for example, expresses disappointment in Kuhn's "choice of 'exemplar' as the meaning of paradigm rather than the more radical combination of ideas and social actions embodied in such notions as 'forms of life' or 'language games'" (see Pinch 1979, 440). This change in Kuhn's view, Pinch suggests, was a return to the traditional "division of labor whereby philosophers of science determine the criteria of good science and sociologists study scientists' use of and deviation from such criteria" (440). Pinch suggests that this return to the traditional division of labor on Kuhn's part was most evident in Kuhn's study of the revolution in physics that gave rise to quantum mechanics. As far as Pinch is concerned, it is a thoroughly internalist history of science (1979, 439).

The move away from a notion of paradigm that "incorporated cognitive and social commitments" led many sociologists to lose interest in Kuhn's work. They were more intrigued by the earlier, radical Kuhn, the one that he took pains to distinguish his later self from.[9]

I return to the topic of Kuhn's internalism in chapter 9. To a large extent, it is because of Kuhn's commitment to internalism that his work is relevant to philosophers of science concerned with developing an epistemology of science. Internalism is one issue that is apt to divide many contemporary sociologists of science and philosophers of science.

My aim in this chapter has been to reconstruct Kuhn's discovery of the concept of paradigm, a concept that is widely regarded as one of his most important contributions to our contemporary understanding of science and scientific inquiry. I have argued that his discovery followed a pattern similar to the pattern of discovery common in science, a pattern that Kuhn himself exposed. Rather than emerging fully formed in *Structure*, the concept of paradigm emerged through a series of phases.

---

[9] In light of this reaction on the part of sociologists of science, it is worth mentioning that the term "paradigm" has gone from being strictly an analysts' term to being an actors' term as well. Although, initially, the term was used only by philosophers, sociologists, and historians of science to describe an aspect of scientific practice, now scientists use the term as well. In fact, two articles in a recent issue of *Science* use the term. In an article titled "Differential Sensitivity to Human Communication in Dogs, Wolves, and Human Infants," Topál *et al.* claim to use "the A-not-B object search paradigm that had been used to demonstrate the influence of communicative cues on human infants' perseverative search errors" (Topál *et al.* 2009, 1,269). And in an article titled "Positive Interactions Promote Public Cooperation," Rand *et al.* claim that "the public goods game is the classic laboratory paradigm for studying collective action problems" (Rand *et al.* 2009, 1,273). These scientists are using "paradigm" in Kuhn's preferred sense. They are identifying or describing the exemplars that they are employing in their research.

In the 1940s and 1950s a number of scholars with Harvard connections working in different disciplines were using the term "paradigm" and they were also coincidentally concerned with the way our perception is shaped by our preconceptions, past experiences, and assumptions. With the publication of *Structure*, paradigms became an integral part of Kuhn's understanding of science. But early criticism revealed that the role of paradigms was unclear. It was only then, as Kuhn responded to criticism, that he finally articulated a precise understanding of the concept of paradigm. In a series of publications in the 1970s, he settled on a conception of a paradigm as a concrete exemplar that functions as a guide to future research. And once he articulated a clear and precise notion of paradigm, he was in a better position to articulate a clear and precise notion of *theory* change. Exemplars, though, continued to play an important role in Kuhn's developed philosophy of science.

There has been a lot of discussion about the extent to which Kuhn's views differ from the views of the positivists, one of the key views that Kuhn was reacting against when he wrote *Structure*. For example, Gattei (2008, chapter 5) and Alexander Bird (2000, 278–80) argue that, to a large extent, Kuhn was unable to move beyond the positivists' assumptions. In closing, I want to draw attention to two significant differences between Kuhn's view and the view of the positivists.[10] First, Kuhn's developed account of theories differs significantly from the positivists' account of theories. According to the positivists, a theory is expressible in a set of sentences. It is from such sets of sentences that predictions are derived. For Kuhn, though, a theory is essentially a set of categories or kind terms. Given the positivist conception of theories, it makes sense to talk of the truth of a theory. A theory is true just in case its constitutive sentences are true. Given Kuhn's conception of a theory, that is, a lexicon, it makes far less sense to describe a theory as true or false. Indeed, given Kuhn's conception of a theory, usefulness and fruitfulness are more appropriate terms of appraisal for theories.

Second, unlike the positivists, Kuhn does not believe that all scientific knowledge is embodied in theories whose content is expressible in sentences. Rather, Kuhn insists that some scientific knowledge is embodied in concrete scientific accomplishments that serve as exemplars for solving hitherto unsolved problems. Kepler's model of the orbit of Mars presented

[10] I recognize, as much contemporary historical scholarship emphasizes, that there is no view that deserves to be called *the* view of the positivists. They were a heterogeneous lot (see, for example, Uebel 2008, 78).

in *The New Astronomy* is a typical exemplar. It provided astronomers with a template for modeling the motion of the other planets. To a large extent, science education is the process by which young scientists learn to work with exemplars, applying familiar solutions to outstanding problems, problems designed to be readily solvable given the conceptual resources supplied by the exemplar.

CHAPTER 4

The epistemic significance of incommensurability

So far, we have examined how Kuhn modified his understanding of scientific revolutions and paradigms and their roles in scientific inquiry and scientific change. Scientific revolutions and paradigms are two of the central concepts in *Structure*. A third key concept introduced in *Structure* concerned Kuhn throughout his career, namely incommensurability. Rather than trace the history of the use of this term, as we did with the concept of "paradigm," my aim here is to distinguish the variety of ways in which Kuhn used the term "incommensurability."

Kuhn regarded the notion of incommensurability as extremely important to understanding scientific change. In fact, in 1990, he claimed that his "own encounter with incommensurability was the first step on the road to *Structure*," adding that "the notion still seems ... the central innovation introduced in the book" (1993/2000, 228). Later in his career, Kuhn devoted more and more energy to the issue of incommensurability.

The concept of incommensurability has generated a lot of interest and caused a lot of controversy both in Kuhn scholarship and in the philosophy of science in general. There is a vast secondary literature on the topic that continues to grow (see, for example, the papers in Hoyningen-Huene and Sankey 2001; and Solar *et al.* 2008). It is widely recognized by both commentators and critics that Kuhn's views on incommensurability developed, though there have been a variety of interpretations of the ways in which his views changed (see, for example, Hoyningen-Huene 1989/1993, 212–18; Sankey 1993; Bird 2000, 291n. 1; Brown 2005; and Andersen *et al.* 2006, chapter 5).

One thing for certain is that Kuhn came to use the term "incommensurable" to describe different phenomena. Sankey and Hoyningen-Huene (2001), for example, distinguish between semantic incommensurability and methodological incommensurability (viii–ix). In this chapter, I discuss four distinct ways in which Kuhn came to use the term "incommensurability." Three of these are presented in *Structure*. Although they have been referred to by a variety of names, I will follow Hacking and refer to them as "topic-incommensurability," "meaning-incommensurability," and dissociation (Hacking 1983, 67). "Topic-incommensurability" corresponds to Sankey and Hoyningen-Huene's methodological incommensurability, and "meaning-incommensurability" corresponds to their semantic incommensurability. "Dissociation" describes the experience of the historian of science as she tries to make sense of some scientific practice of the past that is significantly different from current scientific practices.

In his later writings Kuhn used the term "incommensurable" in yet another way, a way that is somewhat derivative of meaning-incommensurability, but that applies to the conceptual frameworks used in neighboring specialties. Kuhn came to believe that incommensurability aids in the process of specialty formation in science by isolating neighboring research communities from each other, and thus enabling them to develop the conceptual resources appropriate to the phenomena they seek to model. It was only in his later writings that Kuhn attributed this function to incommensurability.

My aim in this chapter is to identify the various ways in which Kuhn thought incommensurability affected science, and to clarify the epistemic significance of incommensurability. Some forms of incommensurability are temporary impediments to our pursuit of knowledge, but at least one form plays a constructive role in advancing scientists' goals.

TOPIC-INCOMMENSURABILITY

It is worth remembering where the concept of incommensurability came from. It has long had meaning in mathematics. The Pythagoreans are credited with discovering that there is no common measure between the hypotenuse and the sides of an isosceles right-angle triangle (see Kuhn 1983/2000, 35). That is, we cannot express both terms in whole numbers. Applied to science, the basic idea is that there is no common measure for evaluating competing theories. This is one of the ways in which Kuhn uses the term "incommensurable" in *Structure*. But even in *Structure*, we will see, Kuhn extended the meaning of the term to cover other phenomena as well.

Kuhn introduces the notion of topic-incommensurability in his discussion of the nature and necessity of scientific revolutions, the first substantive discussion of incommensurability in *Structure*. His intention in that discussion is to explain why "paradigm choice can never be unequivocally settled by logic and experiment alone" (1962a/1996, 94).[1] On Kuhn's interpretation of positivism, the positivists suggest that disputes in science could be resolved in a straightforward manner, by appeal to either logic or experiment. But Kuhn did not believe this was so. He appealed to episodes of theory change in the history of science to support his view. The Copernican revolution, for example, took between six and twelve decades to resolve (see Kuhn 1957, 1). Thus, to some extent, Kuhn invoked the concept of incommensurability in an effort to explain why disputes about which of two competing theories is superior often take years to resolve.

Kuhn claims that "the normal-scientific tradition that emerges from a scientific revolution is not only *incompatible* but often actually *incommensurable* with that which has gone before" (Kuhn 1962a/1996, 103; emphasis added). He gives a sense of what he means in a discussion of successive theories. For example, he notes that "the reception of a new paradigm [that is, a new theory] often necessitates a redefinition of the corresponding science. Some old problems may be relegated to another science or declared entirely 'unscientific'" (103).

Kuhn gives the following example to illustrate his point. In *Principia* Newton "interpreted gravity as an *innate* attraction between every pair of particles of matter" (Kuhn 1962a/1996, 105; emphasis added). But for the scientists trained and still working in the pre-Newtonian mechanico-corpuscular world view, it was imperative to "search for a mechanical explanation of gravity" (105). Given the mechanico-corpuscular theory, this was a legitimate scientific problem. Given Newton's theory, however, the search for a mechanical explanation of gravity was not a scientific problem.

Topic-incommensurability can be clearly illustrated using this example. Imagine two natural philosophers during Newton's time evaluating the two competing theories, the pre-Newtonian mechanico-corpuscular theory and Newton's theory. These two natural philosophers will be led to make different judgments about the two competing theories, depending upon which presuppositions they employ in their evaluations. The advocate of the pre-Newtonian mechanico-corpuscular theory will regard the search for a mechanical explanation as a legitimate scientific problem. As a result, Newton's theory will look deficient. The advocate of Newton's theory will regard gravity as innate, and thus in no need of a mechanical

[1] In this context, Kuhn uses the term "paradigm" to mean theory. Thus, a paradigm choice involves a choice between competing theories. In the previous chapter, we saw that Kuhn came to use the term "paradigm" in a restricted sense, to designate the concrete exemplars that guide scientists in their research. The reader is cautioned to keep this change in mind.

explanation. As a result, this natural philosopher will regard Newton's theory as superior. The two theories are incommensurable because there is no common measure accepted by advocates of both theories for assessing the theories. As a matter of historical fact, many continental Cartesians initially regarded Newton's conception of gravity as a regressive return to occult powers, which they felt had no place in the new science (see, for example, Dear 2001, 164–65). At the same time, many English natural philosophers judged Newton's theory to be superior.

Kuhn also discusses topic-incommensurability in his analysis of the resolution of revolutions. Kuhn claims that "the proponents of competing paradigms will often disagree about the list of problems that any candidate for paradigm [that is, theory] must resolve" (1962a/1996, 148). As a result, he notes, "their standards or definitions of science are not the same" (148). It is crucial to recognize that Kuhn believes that scientific standards are determined, to a large extent, by the problems one seeks to address. If one changes the list of problems one seeks to address, then one changes the standards of evaluation.

The type of rationality that Kuhn is working with here is what is often called instrumental rationality (see Kuhn 1979b/2000, 206; see also Friedman 2001). We judge a choice to be rational or irrational on the basis of whether it advances a goal or set of goals which we take as given. Instrumental rationality makes no judgment about the value of our goals. Two scientists may be led to evaluate competing theories differently if they do not share the same goals. Such differences can create genuine and quite resilient barriers to reaching an agreement about which theory is superior. But Kuhn certainly felt that such barriers could be overcome in time.

Indeed, Kuhn insisted that disputes between advocates of competing theories are generally resolved in a rational manner. But in order to bring such a dispute to a close, additional evidence often needs to be amassed, evidence that makes clear the epistemic superiority of one of the competing theories. At that point, a new consensus emerges, and a new normal scientific tradition begins.

In the discussion of scientific revolutions in chapter 1, we saw that competing theories do not address the same set of topics or problems. The overlap in topics or problems addressed by two competing theories may be quite extensive. In fact, there needs to be quite extensive overlap for the two theories to be deemed *competitors* (see Hoyningen-Huene 1989/1993, 219). Inevitably, though, there will be some problems that the one theory addresses that the other theory does not, and vice versa. And this discrepancy is what often leads two scientists to evaluate competing theories differently. Their standards are not the same. Still, as Kuhn notes, a "lack of a common measure does not make comparison impossible" (Kuhn 1983/2000, 35).[2]

Topic-incommensurability is a rather modest form of incommensurability. But it is as robust a notion as one needs in order to explain what Kuhn sought to explain, that is, why logic and experiment alone cannot settle disputes between advocates of competing theories. Gerald Doppelt (1978) provides one of the clearest explanations of how topic-incommensurability affects scientists. Two scientists who accept or work with different theories are apt to assign greater value to solutions to different problems:

Data which are "anomalous" for the old paradigm [that is, the old theory] but successfully explained by the new enjoy far greater epistemological importance relative to the standards implicit in the new paradigm than they enjoy relative to those implicit in the old paradigm. (Doppelt 1978, 44)

Kuhn notes that very often anomalies are initially just set aside (Kuhn 1962a/1996, 82). There is neither a presumption that they falsify the accepted theory, nor a presumption that they must or can effectively be addressed immediately. But when an anomaly does attract the attention of a sufficient number of scientists, and a solution can be found, those who were moved to take the anomaly seriously enough to develop a means to accommodate it will also likely place great value on a theory that can accommodate the previously anomalous phenomenon. Similarly, those scientists who were not moved to resolve the anomaly are less apt to endorse a new theory, even though it may resolve that particular problem. Hence, topic-incommensurability is a robust enough notion to explain why a research community, when faced with a choice between two competing theories, may not be able to quickly resolve differences about which of the competing theories is superior.

Incidentally, Hacking notes that in 1960 most philosophers would have thought that successive theories in a field subsume the theories they replace. That is, the new theories retain *all* the successes of their predecessors, and "cover a wider range of phenomena and predictions" than their predecessors

---

[2] Brown (1983) notes that critics of both Kuhn's analysis and Feyerabend's analysis of incommensurability "have taken incommensurable theories to be theories which cannot be compared in a rational manner" (3). Both Kuhn and Feyerabend reject this interpretation (Brown 2005, 157; see also Hoyningen-Huene 1989/1993, 218–22). Given Feyerabend's diagram of incommensurable theories in his contribution to *Criticism and Growth of Knowledge*, however, it is easy to see why some were led to think that he thought such theories cannot be rationally compared (see Feyerabend 1970/1972, 220, figure 2).

(see Hacking 1983, 67–69). Now, though, that view seems untenable and topic-incommensurability seems undeniable (see Hacking 1983, 69).

As a matter of fact, Karl Popper (1975/1998) continued to endorse this view that Kuhn was reacting against, the view that Hacking says was taken for granted in 1960. According to Popper, "a new theory, however revolutionary, must always be able to explain fully the success of its predecessor. In all those cases in which its predecessor was successful, it must yield results at least as good as those of its predecessor and, if possible, better results" (1975/1998, 291).

It is worth highlighting the key difference between the type of disparity in evaluation that concerns Kuhn in his discussion of topic-incommensurability and the type of disparity in evaluation that he discusses in "Objectivity, Value Judgment, and Theory Choice" (1977c). In this paper, Kuhn argues that two scientists can appeal to the same set of values – simplicity, breadth of scope, accuracy, consistency, and fruitfulness – in evaluating competing theories and be led to different conclusions about which theory is superior. The discrepancies in judgments may result from the fact that the scientists weigh the various criteria differently, or they differ in how they understand the various criteria. Simplicity, for example, can mean different things in different contexts. And competing theories can be simple in different respects. In his discussion of topic-incommensurability, on the other hand, Kuhn is claiming that two scientists may be led to disagree on their evaluation of competing theories because they are concerned with different sets of problems, a consequence of the fact that different theories address different problems.

It is the notion of a common measure that is of most importance for explaining what Kuhn sought to explain when he initially invoked the concept of incommensurability. He was seeking to explain why disputes between scientists cannot be resolved by appeal to logic and experiment alone. Topic-incommensurability explains why a change of theory is often a protracted affair. Contrary to what some of his critics suggest, Kuhn never conceived of incommensurability as an insurmountable barrier to theory change or theory evaluation. Nor was the concept intended to imply that the resolution of such disputes was either irrational or non-rational.[3]

---

[3] The concept of incommensurability has also found its way into political philosophy. In "Two Concepts of Liberty," Isaiah Berlin describes the various kinds of freedom that people can enjoy as incommensurable with each other (Berlin 2002, 177n. 1). He also came to describe the various goods that people pursue as incommensurable. Berlin's notion is similar to topic-incommensurability. It emphasizes that there are a number of potentially conflicting values that make evaluations of alternative ways of life, or theories (in the case of science), difficult.

MEANING-INCOMMENSURABILITY

Consider the second way the term "incommensurability" is used in *Structure*, what has come to be called "meaning-incommensurability." According to Kuhn, when a new theory replaces an older theory in a particular field, "old terms, concepts, and experiments fall into new relationships one with the other" (1962a/1996, 149). As a result, "communication across the revolutionary divide is inevitably partial" (149). To illustrate his point, Kuhn discusses the change in meaning that Copernicus introduced with his new, heliocentric theory. According to Kuhn, "part of what [Copernicus' predecessors] meant by 'earth' was a fixed position. Their earth ... could not be moved" (1962a/1996, 149). Consequently, "Copernicus' innovation ... was a whole new way of regarding the problems of physics and astronomy, one that necessarily changed the meaning of both 'earth' and 'motion'" (149–50).

In another context, Kuhn discusses the example of the change in the meaning of "mass" from Newtonian physics to Einsteinian physics.[4] Given that the central terms of scientific theories are defined in relation to each other in a holistic manner, such changes of meaning can make it very difficult for scientists to effectively compare competing theories. Difficulties in communication between advocates of competing theories are thus to be expected.

Most philosophers concerned with the issue of incommensurability have been concerned with meaning-incommensurability (see Bird 2000, 150–51). Doppelt (1978) argues that the interest in and focus on meaning-incommensurability is due to Kuhn's early critics, especially Dudley Shapere (1966/1981) and Israel Scheffler (1967). Sankey's book-length study of incommensurability is quite typical of the contemporary literature on incommensurability. In his book, Sankey is exclusively concerned with meaning-incommensurability (1994, 1).[5] He does not even acknowledge that there are different forms of incommensurability. Thus, Sankey totally neglects topic-incommensurability. Given his myopic focus on meaning-incommensurability Sankey is led to conclude that "since so few of the

---

[4] Hacking discusses this example in his analysis of meaning-incommensurability (1983).

[5] Meaning-incommensurability is the focus of many commentators, including Collier (1984), Chen (1990), and Sankey (1991, 1994). Even some historians of science who discuss incommensurability focus on meaning-incommensurability (see Buchwald and Smith 2001). In a recent comparison of Feyerabend's and Kuhn's notions of incommensurability, Oberheim (2005) suggests that Feyerabend was concerned with meaning-incommensurability (386). Further, Oberheim notes that whereas Feyerabend was led to the concept through the literature on the psychology of perception, Kuhn was led to the concept through his study of the history of science (385).

radical claims associated with the incommensurability thesis are warranted by the phenomenon of conceptual change in science, it is not clear that there is anything left for the word 'incommensurability' to stand for" (221).

Sankey is correct in his assessment of the threat posed by incommensurability provided one only considers meaning-incommensurability. Meaning-incommensurability is far less threatening than some have been led to believe. The fact that key terms do not have the same meaning in two competing theories does not entail that the theories cannot be compared. Nor does it entail that there is no rational basis for assessing the theories. But meaning-incommensurability is only one form of incommensurability. And some of the other forms do have interesting epistemic implications. Indeed, topic-incommensurability is an important concern for the epistemology of science, and one that has received less attention than it deserves, in part, because Kuhn's attention turned elsewhere.

In the later part of his career Kuhn seems to have been preoccupied with the issue of meaning-incommensurability. In fact, Kuhn notes that after the publication of *Structure*, he "increasingly identified incommensurability with difference of meaning" (Kuhn 1993/2000, 237; see also Brown 2005, 152). But the issue of meaning change is distinct from the issue of topic-incommensurability, for changes of meaning do not necessarily involve changes in standards. At times, it seems that Kuhn recognized that he was drifting into a new topic in his discussions of meaning-incommensurability, for he notes that:

[A]pplied to the conceptual vocabulary deployed in and around a scientific theory, the term "incommensurability" functions metaphorically. The phrase "no common measure" becomes "no common language." (Kuhn 1983/2000, 36)

Hence, strictly speaking, meaning-incommensurability does not involve a lack of shared standards. In fact, in a discussion of what he called "meaning incommensurability," Kuhn notes that "what ... is at issue is not significant comparability but rather the shaping of cognition by language" (1983/2000, 55). Competing theories are like different languages in that they "impose different structures on the world" (52).[6]

Despite the fact that meaning-incommensurability is not concerned with the absence of shared standards, meaning-incommensurability can make the evaluation of competing theories difficult. When two scientists

---

[6] Alexander Bird (2000) has suggested that Kuhn's engagement with issues in the philosophy of language were, for the most part, misguided and unfruitful. I agree with Bird. And it seems that Kuhn was aware that some of his remarks on meaning-incommensurability were muddled, for he admits to confusing the notions of language learning and translation in his discussions of meaning-incommensurability (see Kuhn 1993/2000, 238).

have different ideas and expectations about the referent or extension of a kind term as they do when they work with competing theories, "communication is … jeopardized" (Kuhn 1993/2000, 231). As a consequence, Kuhn claims, it can be difficult to rationally adjudicate between two competing theories (231). Hence, the fact that competing theories use the same terms in different ways does contribute to making the resolution of disputes in science protracted. But we should not confuse differences of meaning that make it difficult for scientists to resolve their disagreements with differences of standards.

In Kuhn's later writings, he came to describe meaning-incommensurability as *local* incommensurability. During an episode of theory change, the terms whose meanings change from one theory to its successor often are relatively few and can be locally contained. That is, they affect only a small part of the scientific lexicon (in this regard, see Andersen *et al.* 2006, 105–08). As a result, Kuhn claims, "the terms that preserve their meanings across a theory change provide sufficient basis for the discussion of differences and for comparisons relevant to theory change" (Kuhn 1983/2000, 36). Consequently, Kuhn makes it clear that he did not believe that meaning-incommensurability undermined the possibility of comparing competing theories (see 1983/2000, 34 and 36).

Andersen *et al.* (2006) provide a valuable analysis of how we can understand the nature and dynamics of local incommensurability employing a frame model of concepts that was developed in the cognitive sciences (see especially chapter 5 there). Importantly, this account of concepts has many affinities to Kuhn's own understanding of concepts. Andersen *et al.* take the fact that the frame model of concepts is now widely accepted as a vindication of Kuhn's own account.

DISSOCIATION

The third type of incommensurability that Kuhn discusses in *Structure* is what Hacking refers to as "dissociation." Kuhn claims that "the proponents of competing paradigms [that is, theories] practice their trade in different worlds" (1962a/1996, 150). When he wrote *Structure*, he described this as "the most fundamental aspect of the incommensurability of competing paradigms" (150). And the notion of scientists working in different worlds has been one of the most elusive and disturbing aspects of Kuhn's view.

In an effort to illustrate the nature of dissociation Hacking discusses the example of Paracelsus' medical writings. According to Hacking, "Paracelsus's discourse is incommensurable with ours … because there is no way to match what he wanted to say against anything we want to say" (1983, 71). Elaborating, Hacking claims that "one can start talking [Paracelsus'] way only if one becomes alienated or *dissociated* from the thought of our own time" (71; emphasis added).

In a recent article, Ipek Demir (2008) distinguishes between the incommensurability that "*scientists* encounter during revolutionary periods," and the incommensurability that *analysts* of science, for example historians of science, encounter "when they engage in the representation of science from earlier periods" (133). Dissociation describes the *analyst's* experience. The historian of science is apt to experience dissociation as she seeks to understand past scientific practices and theories. The scientist who happens to live through a change of theory in her specialty, on the other hand, experiences something quite different.

Interestingly, Kuhn notes that he first became aware of incommensurability when he was working as an *analyst* rather than when he was working as a scientist (Kuhn 1991a/2000, 91). It was "from attempts to understand apparently nonsensical passages encountered in old scientific texts" that he was led to this notion of incommensurability (91). On numerous occasions, Kuhn recounts the trouble he had trying to make sense of Aristotle's physical theory (see Kuhn 1987/2000, 15–17). His familiarity with Newton's physics interfered with his understanding of Aristotle's physical theory (see Kuhn 1977b, xi). It was only when he began to see that Aristotle's concerns were not the same as Newton's concerns that he began to see that by Aristotle's own standards Aristotle was a good physicist.

SPECIALIZATION AND INCOMMENSURABILITY

In chapter 7, I examine Kuhn's account of specialty formation in science in detail. For now, though, it is worth noting that Kuhn came to believe that some crises in science are resolved, not by the replacement of one theory by another, but by the creation of a new scientific specialty. Each field, the parent field and the new specialty, develops its own lexicon (Kuhn 1991a/2000, 98).[7] He came to describe the two lexicons as *incommensurable* with each other (98).

---

[7] Kuhn suggests that Biagioli's (1990) paper, "The Anthropology of Incommensurability," helped him see this (see Kuhn 1991a/2000, 97). Biagioli describes the incommensurability between competing paradigms as playing "an important role in the process of scientific change and *paradigm-speciation*" (Biagioli 1990, 183; emphasis added).

Kuhn suggests that the resulting incommensurability that arises between neighboring specialties is inescapable. The creation of a new scientific specialty, and the conceptual barriers it creates, are sometimes the only way we can make progress in our pursuit of scientific knowledge (98). The conceptual resources available in the parent field prove inadequate for the range of phenomena scientists seek to model. The solution is to create two fields, each concerned with a narrower range of phenomena. It is in this respect, Kuhn claims, that specialization is similar to biological evolution (see Kuhn 1991a/2000, 97). Scientists are developing more specialized niches in which to work. In chapter 7, I discuss in some detail two examples of this process, the creation of the field of endocrinology and the creation of the field of virology.

Kuhn claims that this kind of incommensurability that occurs between neighboring scientific specialties plays an important role "as an isolating mechanism" (1991a/2000, 99; also 1992/2000, 120). Just as physical barriers, like mountains or wide waterways, aid with speciation in the biological world, the isolation that results from the meaning-incommensurability that develops between specialties facilitates conceptual development. Such isolation allows each group to develop a lexicon suited to the phenomena that concern them and their research. As a result, *collectively*, the various scientific specialty communities are better able to realize their goals. Kuhn puts the point in the following way: "it is the specialization consequent on lexical diversity that permits the sciences, viewed collectively, to solve the puzzles posed by a wider range of natural phenomena than a lexically homogenous science could achieve" (1991a/2000, 99). Efforts to unify science, especially efforts to create a unified scientific lexicon, can be an impediment to science. And the incommensurability that emerges between specialties aids scientists by foiling such efforts at unification.

The type of incommensurability that develops between neighboring scientific specialties is akin to meaning-incommensurability. Scientists working in neighboring specialties are often impeded in effective communication across specialty lines because they attach different meanings to the same terms. But unlike the phenomenon Kuhn refers to as "meaning-incommensurability," which is often a barrier to progress in science, the differences in meaning that divide neighboring specialties serve to advance the goals of science. They do this by allowing each research community to develop concepts that serve their *local* goals.

Consider, for illustrative purposes, the concept "species." It is widely recognized that there are a number of incommensurable species concepts (see Ereshefsky 1998). There is an interbreeding concept, a phylogenetic concept, and an ecological concept (see Ereshefsky 1992; 1998, 105). Importantly, these different concepts "carve the tree of life in different ways" (105). Thus, "many interbreeding species fail to be phylogenetic species, and many phylogenetic species fail to be interbreeding species" (105). Each species concept is suited to a different set of research interests. And scientists working in different sub-fields of biology tend to work with one conception only. Paleontologists, for example, prefer the phylogenetic concept, whereas population geneticists work with the interbreeding conception.[8] Were biologists to try to overcome the incommensurabilities between their sub-fields, they would likely be frustrated in the pursuit of their research goals. There is no need to develop a single unified species concept. In fact, to attempt to develop such a concept might be detrimental to science.

In Kuhn's later writings, generally, when he discusses incommensurability it is either meaning-incommensurability or the incommensurability between the lexicons in neighboring specialties. But he continued to regard all three forms of incommensurability that he introduced in *Structure* as important.

In summary, I would like to briefly explain how the four forms of incommensurability identified by Kuhn differ with respect to their epistemic significance.

Dissociation is of no real concern to scientists. Rather, it describes the experience of the analyst, the sociologist, or historian of science trying to understand earlier theories which no longer have much in common with contemporary scientific theories. Given that earlier scientists had concerns that are often far removed from the concerns of contemporary scientists, it can be challenging for the sociologist or historian to develop an adequate understanding of the views and concerns of earlier scientists (see Cohen 1974, for a discussion of some of the challenges). But scientists need not understand their past in order to be effective scientists.

Meaning-incommensurability does affect scientists. It can cause confusions between proponents of competing theories. For example, the different meanings of "mass" in the theories of Einstein and Newton could lead to confusion between advocates of each theory, that is, until each understood how their use of the term differs from the others' use. Thus, the real threat posed by meaning-incommensurability is misunderstanding.

---

[8] I thank Marc Ereshefsky for guidance on this issue, especially with identifying which concept serves which sub-field.

But the effects of misunderstanding in scientific disputes should not be underestimated.

The related concept of meaning-incommensurability that arises between different specialty communities is similar in that it can lead to persistent misunderstandings for those who try to speak across specialty lines. But Kuhn insists that these misunderstandings serve a positive epistemic function. They serve to isolate the two groups from each other, which in turn allows each group to develop the conceptual resources that each needs. Without such isolation, the groups may find they are less capable of developing the concepts, instruments, and practices suited to their objects of study.

Finally, topic-incommensurability helps explain why logic and experiment do not enable scientists to readily reach agreement about which of two competing theories is superior. Because scientists who accept different theories are often concerned with different topics or problems, they will not necessarily agree about which of two competing theories is superior, even if they have access to the same body of data. They lack a common measure by which to evaluate the competing theories. And it is because of topic-incommensurability that disputes between advocates of competing theories are often rather protracted affairs. For example, it is due, at least in part, to topic-incommensurability that the Copernican revolution in astronomy was such a drawn-out affair. Concerned with different scientific problems, the advocates of each of the competing theories appealed to different standards and were led to disagree about which theory was superior. Only as each theory was refined and new data were collected were astronomers able to reach an agreement about which theory is superior. The competing theories must be developed before one theory shows itself to be unequivocally superior to the other. Indeed, if Kuhn's account of theory change is correct, then, in general, it is only as those working in a field come to agree about what problems are most important that they will be able to reach a consensus about which theory is superior.

Kuhn's evolutionary epistemology

Kuhn's epistemology of science is an evolutionary epistemology. Critics and commentators alike have generally either ignored or misunderstood this dimension of his project. My aim in Part II is to rectify this situation. I aim to show that understanding Kuhn's evolutionary perspective on epistemology is the key to understanding his epistemology of science. Such a perspective, I argue, is at odds with the perspective most philosophers bring to their study of science. Kuhn's approach to evolutionary epistemology requires a radical shift in perspective. Indeed, this is one reason why Kuhn is so frequently misunderstood.

No doubt, part of the reason Kuhn's evolutionary epistemology is misunderstood is the fact that the term "evolutionary epistemology" covers a wide range of projects, with very different aims. Perhaps the most popular approach to evolutionary epistemology is that which seeks to explain our many true beliefs in terms of the evolutionary advantages accrued to the sorts of creatures who have developed the means to acquire the beliefs. For example, one might explain our many true beliefs about the visual aspects of the world in terms of our capacity to see. This is the sort of project that Donald Campbell (1974) was pursuing. And, for a time, the project was quite popular. It seems to be a common-sense way to understand the project of naturalizing epistemology. Our best scientific theories, including our knowledge of biological evolution, provide insight into understanding our epistemic successes. This sort of project is also compatible with reliabilist theories of justification, according to which a belief is justified insofar as it was acquired by some reliable means.[1]

---

[1] Campbell (1974) provides a seven-page bibliography on evolutionary epistemology, much of it historical, identifying many early evolutionary explanations and analogies offered by philosophers, psychologists, and scientists seeking to explain how we acquire knowledge. His preferred model is a "blind-variation-and-selective-retention process."

Alvin Goldman's (1986) *Epistemology and Cognition* is the classic source on reliabilism in epistemology.

There are, however, limitations to such an approach to evolutionary epistemology. It is quite plausible to believe that our evolutionary history can explain why it is that people can discern certain colors and smells, for example. After all, we can imagine a variety of advantages that would have accrued to our early ancestors as they developed such abilities. And evolutionary theory might offer some insight into why certain basic patterns of reasoning are pervasive. But it is quite implausible to believe that there is an evolutionary explanation underlying physicists developing or accepting a particular theory of the atom, for example. We have accepted the theories we have in part because of the sources of information we have acquired and refined throughout our evolutionary history. But it is hard to believe that such theories can be explained in terms of the evolutionary fitness they have afforded us. Hence, this sort of approach to evolutionary epistemology does not seem suited to scientific knowledge, especially theoretical knowledge.[2]

Philosophers of science have developed alternative approaches to evolutionary epistemology, approaches more suited to understanding science and scientific knowledge. Popper, for example, compares the testing of competing theories in a research community to the selection of the fittest variations in a biological population. The testing that our theories undergo, he claims, is like the challenges that species must overcome if they are not to be driven to extinction (see, for example, Popper 1972). Popper's own version of evolutionary epistemology, though, has been widely criticized, even by those who are otherwise sympathetic to Popper's philosophy of science (see, for example, Baigrie 1988).

More recently, in *Science as a Process*, David Hull (1988) has developed an evolutionary epistemology of science. On Hull's account, the various institutions that constitute the environment in which competing hypotheses encounter each other are organized in a manner such that weaker hypotheses are weeded out. Hull, though, does not believe that the epistemic effectiveness of science is a consequence of the fact that the

---

[2] The literature on evolutionary epistemology is vast. In fact, in an article published in 1986, Michael Bradie provides a six-page bibliography of literature on evolutionary epistemology (see Bradie 1986). Consequently, my discussion will be quite selective. This particular approach to evolutionary epistemology that seeks to explain our propensity to adopt true beliefs or our tendency to reason correctly in terms of selection pressures has been criticized for a number of reasons. Stephen Downes (2000), for example, argues that a number of "versions of the view that mechanisms of true belief generation arise out of natural selection ... fail to establish a connection between truth and natural selection" (425). And Richard Feldman (1988) argues that "even if it is advantageous to use rational strategies, it does not follow that we actually use them; and ... natural selection need not favor only or even primarily reliable belief-forming strategies" (218). See also Stephen Stich's (1990) *Fragmentation of Reason*, chapter 3.

institutions constitutive of science are well designed with great foresight, aimed at eliminating weaker theories. Humans are not so clever or capable of designing these institutions, knowing well in advance the ends they will aid us in achieving. Rather, Hull believes that such institutions have developed gradually over time. Through trial and error, the institutions constitutive of science have been modified in ways to make them more effective at advancing scientists' epistemic goals.[3]

According to Hull, the key to the success of science lies in the fact that the constitutive institutions, in conjunction with the reward structure in science, encourage certain types of behavior in scientists, behaviors that, luckily, are conducive to aiding scientists in realizing the goals of science. For example, scientists seek peer recognition. That is a key source of motivation for researchers. Consequently, in their efforts to secure the type of positive peer recognition they desire, they are forced to produce good research that will stand up to the scrutiny of their peers, and that will be deemed useful to advancing their peers' research goals. In this way, scientists are driven to produce good research, which in turn advances the institutional goals of science. Hull thus offers a functional explanation of peer review.[4]

Such functional explanations of the social institutions of science were developed earlier by the sociologist of science Robert K. Merton (see Merton 1959). In his studies of priority disputes and multiple discoveries in science Merton aims to show how the institutions of science, like the peer review process, ensure that science functions properly. But even

---

[3] Friedrich Hayek (1960) also argues that the institutions constitutive of modern societies have evolved gradually and without design. Further, he believes they embody much of the knowledge that we take for granted. See especially chapters 2 and 3 in *Constitution of Liberty*.

  Hull's cynicism about the ability of planners to effectively direct science is shared by Popper. Indeed, the mistrust of politicians and planners to direct society in general is the central theme of Popper's (1946/1950) *The Open Society and Its Enemies*.

[4] The various evolutionary epistemologies developed by philosophers of science have also been subject to criticism. Paul Thagard's (1980) criticism is quite typical, insisting that "the similarities between biological and scientific development are superficial" (187). Similar complaints are raised by L. J. Cohen (1973). Hull (1974), though, believes that many criticisms of evolutionary epistemologies of science are based on a mistaken view about biological evolution. For example, Hull criticizes Cohen for failing to realize that the unit of biological evolution is not the individual organism, but rather a population of organisms (Hull 1974, 334). Further, Hull objects to the common assumption raised by critics of evolutionary epistemologies of science that cultural evolution – the sort of evolution that would be relevant to explaining the success of science – is Lamarckian not Darwinian (see Hull 1988, 452). Hull is concerned that most of the philosophers who raise this point do not understand Lamarck's view. Specifically, these critics fail to realize that "in order for a form of inheritance to count as Lamarckian ... the acquired characteristic must be inherited. Nongenetic transmission is not good enough" (1988, 453). Lamarck's theory is thus no less genetic than Darwin's.

though the institutions generally operate in ways that aid scientists in realizing their research goals, habits and practices sometimes emerge that are dysfunctional. For example, when two scientists engage in a priority dispute, and their claims were made a mere two weeks apart, the system of peer recognition which honors only the first to make the discovery is not fulfilling its function.

To a large extent, Kuhn accepts this general approach to evolutionary epistemology. That is, like Hull and Merton, he believes that many of the institutions and practices of science have a function which aids in advancing the goals of science. Indeed, Kuhn uses the term "function" intending it to mean just what Hull and Merton mean. For example, Kuhn discusses the *function* served by the Whig histories of science that are written for science textbooks (see Kuhn 1962a/1996, 137). These histories misinform young scientists by distorting what really happened for the purpose of reinforcing the view that scientific knowledge is cumulative, and that the previous research in a field was leading to the currently accepted view or theory. In reality, though, as Kuhn notes, earlier scientists often had very different interests and goals from those of contemporary scientists. Early modern physicists generally believed that God created the world, for example. Contemporary textbook presentations of Boyle's, Hooke's, and Newton's work seldom mention this aspect of their lives. It is deemed both irrelevant and "unscientific," despite the fact that these early modern scientists did not see things this way.

As Kuhn developed his epistemology of science, he saw more and more similarities between biological evolution and scientific change. Consequently, as he developed his epistemology of science it became a more thoroughly evolutionary epistemology of science.

I begin Part II by tracing an important development in Kuhn's thinking about science, a development that has important implications for understanding his evolutionary epistemology. Kuhn was one of the key philosophers of science who initiated the *historical turn* in philosophy of science in the early 1960s. Later, though, he changed his attitude about the relevance and role of the history of science to philosophy of science. He came to adopt what he later called a *historical perspective*. This historical perspective, or developmental view, as he sometimes called it, is an evolutionary perspective on science. Whereas the historical *turn* in the early 1960s led him and others to look to the history of science for data in their efforts to construct an adequate philosophy of science, the historical *perspective* provides a new way to understand science and scientific change. The historical perspective enables us to see science as in process, and to see that scientists work within a tradition of accepted beliefs. This is a radically different perspective on science from the perspective that had previously informed philosophy of science. Consequently, adopting a historical perspective significantly changes our understanding of science and scientific knowledge.

Given the historical perspective that Kuhn adopts, it can be and has been challenging to figure out what Kuhn's answers are to some of the classic questions that concerned philosophers of science. Some of the traditional questions no longer make much sense after we have adopted a historical perspective.

Importantly, the historical perspective also leads us to rethink the role that truth plays in explaining the success of science. Kuhn believes we can make better sense of scientific inquiry and the success of science if we see scientific inquiry as pushed from behind, rather than seeing science as aiming toward a fixed goal set by nature. This is not to say that the world does not constrain our theorizing. Kuhn certainly believes it does. Rather, what he wants us to see is that the scope of our theories is not determined by nature in advance of our inquiring about it. Importantly, this is one respect in which Kuhn's approach to the study of science is more consonant with sociological studies of science. It is an approach that philosophers could benefit from.

Kuhn began to think of his epistemology of science in evolutionary terms at the end of *Structure*. There, Kuhn briefly explains why he seldom mentions truth in a book devoted to scientific knowledge. He compares scientific change to evolutionary change, arguing that just as evolution is not driven toward a goal set in advance, science is not aiming at a goal set by nature in advance.

Philosophers have generally taken for granted that science does have a goal set by nature in advance. As a result, the success of science is generally understood to be a measure of how we are doing with respect to this goal. Kuhn, though, believes that such an image of science is both misleading and not very illuminating. This insight was central to Kuhn's epistemology of science in *Structure* and he continued to regard it as important to the end. Surprisingly, though, it did not attract much attention until recently (see Bird 2000; Renzi 2009).

When Kuhn wrote *Structure*, he already believed that the truth did not explain much about the success of science. But he did not yet have a positive answer to the question of what could explain the success of science. Consequently, at that stage, he was really only prepared to argue that the goals of science are *not* set by nature in advance. Later, though, when he adopted the historical perspective, he realized that specialization can account for some of the aspects of scientific inquiry that philosophers had previously sought to account for by appealing to the truth. Most importantly, he saw how specialization serves to advance the epistemic goals of science, by allowing scientists to develop more precise conceptual tools for modeling the parts of nature they seek to understand. Hence, in his later writings, Kuhn began to develop a positive answer to the question to which he could offer only a negative answer when he wrote *Structure*.

CHAPTER 5

Kuhn's historical perspective

Kuhn was part of the vanguard that ushered in the historical turn in philosophy of science which looked to the history of science as a source of data for developing a philosophy of science. This was a monumental change in philosophy of science, marking, if not causing, the demise of positivism. The historical turn had a wider impact, contributing to important developments in the sociology of science, including the rise of the Strong Programme in the Sociology of Scientific Knowledge. In this respect, Steve Fuller (2000) is correct to claim that with the writing of *Structure* Kuhn unleashed a series of events that were no part of his intentions.

As *Structure* was subjected to criticism Kuhn changed his view about the relevance of the history of science to the philosophy of science. He came to believe that the key insight that *philosophers* could gain from the history of science was a particular perspective on science, a historical or developmental *perspective*. In this chapter, I aim to both trace the path that led Kuhn to this change of view and to clarify what it is that the historical perspective offers us.

According to Kuhn, the historical perspective helps us see that scientists (1) always work within a tradition, beginning their inquiries with a set of beliefs inherited from their predecessors, and (2) are concerned with the evaluation of *changes* of belief rather than with the evaluations of *belief*.

Central to Kuhn's view, especially as it was developed in the latter part of his life, is a radical proposal about the *end* of scientific inquiry. Traditionally philosophers have uncritically assumed that science aims at the truth, and the increasing accuracy achieved in science is taken as evidence that we are getting ever closer to the truth. According to Kuhn, progress in science is less a result of our getting increasingly closer to the truth than it is a result of our developing specialty communities, research communities that develop theories specifically designed to model a narrow range of phenomena. Hence, we should see science as a process of increasing specialization. This particular dimension of Kuhn's philosophy of science has generally been either neglected or misunderstood. In certain respects, specialization fills the part played by truth in traditional philosophical accounts of science.

KUHN AND THE HISTORICAL TURN

The philosophers of science who preceded Kuhn, the positivists and Karl Popper, were primarily concerned with the *logic* of science (see Kuhn 1970a/1977, 288). They explicitly eschewed both the psychology of science and the history of science, believing that these disciplines are irrelevant to answering questions about confirmation, which is the proper concern of philosophers of science (see Butts 2000, 195–96; Brown 2005, 159–60).[1] Popper, for example, distinguishes between the psychology of knowledge and the logic of knowledge, arguing that the widely held but mistaken "belief in inductive logic is largely due to a confusion of psychological problems with epistemological ones" (1959, 30). Similarly, Hans Reichenbach insists that "there is a great difference between the system of logical interconnections of thought and the actual way in which thinking processes are performed [in science]" (1938/2006, 5). Only the former, Reichenbach suggests, are of interest to the philosopher of science.

Moreover, Popper's and the positivists' interest in the history of science was for *illustrative* purposes only. A well-chosen historical example could illustrate a key logical point. For example, in *Experience and Prediction*, Reichenbach discusses Michelson's experiment that is taken to show "the equality of the velocity of light in different directions" (Reichenbach 1938/2006, 84). Reichenbach uses this historical example to illustrate the difference between the types of claims we can know with certainty and the types of claims we can know with only some degree of probability, a distinction he believes transcends the contingencies of history (84–85). Reichenbach also discusses an experiment of Lavoisier's as an example of a *crucial experiment* "in favor of the oxidation theory of combustion"

---

[1] Although Kuhn criticizes both Popper and the positivists for focusing too narrowly on the logic of science, he did regard Popper as an ally, "united in opposition to a number of the most characteristic theses of classical positivism" (Kuhn 1970a/1977, 267). He notes, for example, that they "both emphasize ... the intimate and inevitable entanglement of scientific observation with scientific theory; [they are both] ... skeptical of efforts to produce any neutral observation language; and [they both] insist that scientists may properly aim to invent theories that *explain* observed phenomena" (267). In a footnote, Kuhn adds that "both insist that adherence to a tradition has an essential role in scientific development" (Kuhn 1970a/1977, 267–68n. 4; see also Fuller 2004, 21).

(1938/2006, 388–89). Popper also discusses Lavoisier's experiment in order to illustrate his *concern* about the logic of confirmation. Given Popper's falsificationism, there are no confirming crucial experiments, only falsifying ones (Popper 1959, 78n. 1). Hence, as far as Popper is concerned, though Lavoisier's experiment refutes the phlogiston theory, it "[cannot] establish the oxygen theory of combustion" (Popper 1963, 220).

In the first chapter of *Structure* Kuhn proposes to use the history of science in a different way. He proposes to use the history of science as the source of data from which to develop a philosophy of science (see Kuhn 1962a/1996, chapter 1; 1976/1977, 4; 1991a/2000, 95; 1992/2000, 107).[2] Kuhn was not alone in pursing the philosophy of science this way. Paul Feyerabend, Russell Hanson, Mary Hesse, and others were pursuing a similar strategy. Kuhn explains that he and the others "turned to history [to build] a philosophy of science on observations of scientific life, the historical record providing [our] data" (1992/2000, 107).[3]

Kuhn, though, did not think that the history of science was *merely* a means for testing our hypotheses about the nature of science and scientific change. Rather, he believed that the history of science might also "prove to be a particularly consequential source of problems and of insights" (1976/1977, 4). That is, were philosophers of science to begin with a careful study of the history of science, they might be led to ask new and more fruitful questions about the nature of science and scientific knowledge.

Kuhn suggests that similar insights to those he learned from the history of science about the nature of science could also be gained from learning contemporary science (see Kuhn 1976/1977, 13). The key is that in order to develop an adequate philosophy of science philosophers must "more closely acquaint [themselves] with science" (Kuhn 1976/1977, 13). History is but one means to this end.[4]

---

[2] It is worth noting that Kuhn believed that history of science is an autonomous discipline, with goals distinct from those of philosophy of science (1976/1977, 5). Indeed, he claims to have encountered difficulties in attempting "to draw the two fields closer together" (1976/1977, 4). Kuhn grants, though, that one person can work in both disciplines, as he in fact did. What he denies is that one can work in both disciplines simultaneously (5). Incidentally, Mary Hesse (1976), another pioneer of the historical school, also emphasizes the importance of recognizing the autonomy of history of science, and in particular recognizing that history of science has goals distinct from the goals of philosophy of science.

[3] Robert Butts notes that this idea that "proper philosophy of science can only come from a close study of the history of science" was also held long before by both William Whewell and Pierre Duhem (Butts 2000, 200). But, as Butts notes, Whewell and Duhem were "ahead of their time."

[4] Ron Giere also believes that insofar as the history of science is relevant to philosophy of science it is *as science, not history*. That is, just as a study of contemporary science provides insight into the nature of science so too does a study of early modern or nineteenth-century science (see Giere 1973, 295).

Kuhn was quite surprised at what he found in his study of the history of science (Kuhn 1962a/1996, vii). Two lessons stand out as especially important. First, he found that "methodological directives, by themselves [were insufficient] to dictate a unique substantive conclusion to many sorts of scientific problems" (1962a/1996, 3). Second, he discovered that the results of observations were not "mere facts, independent of existing belief and theory" (1992/2000, 108; see also 1962a/1996, 7). Rather, "the supposed facts of observations turned out to be pliable" (1992/2000, 107–08). Further, Kuhn noted that "producing [facts has often] required apparatus which itself depended on theory, [and] often on the theory that the experiments were supposed to test" (108). The history of science thus taught Kuhn that (1) theory choice is underdetermined, and (2) observations are theory-laden. Given the malleable nature of data and the underdetermination of theory choice, Kuhn realized that the data could not play the role he had *assumed* they played in *resolving* disputes in science.

Kuhn explains that before he looked carefully at the history of science he had uncritically accepted a particular view of scientific knowledge, one that privileged observation, treating it as the *foundation* of scientific knowledge. According to this view:

- "science proceeds from facts given by observation"
- "those facts are objective in the sense that they are interpersonal"
- facts "are prior to the scientific laws and theories for which they provide a foundation"
- "to find [laws, theories, and explanations] one must interpret the facts"
- when scientists are confronted with a choice between competing theories "observed facts … provide a court of final appeal" (Kuhn 1992/2000, 107).

By the 1960s, many philosophers found this view of science objectionable. And, like Kuhn, many came to believe that the history of science would be a valuable resource for developing an alternative philosophy of science (see Kuhn 1968/1977, 121; Butts 2000, 196n. 7).[5]

---

[5] Alan Richardson (2007) has recently argued that, contrary to what Kuhn implies, the view of science that Kuhn was reacting against in *Structure* should not be identified with the view of the logical positivists working in the 1950s (160). Indeed, Richardson claims that Kuhn was remarkably ignorant about the state of positivism in the late 1950s and early 1960s. Kuhn was certainly unaware of Carnap's later work, the work that has struck a number of contemporary philosophers as similar in important respects to Kuhn's view in *Structure* (Kuhn 2000b, 305–06; see also Reisch 1991; Fuller 2000, 391; Friedman 2001). And Kuhn admits as much (see Kuhn 1993/2000, 227). Indeed, it seems that the principal source from which Kuhn acquired *his* understanding of positivism is Reichenbach's *Experience and Prediction* (see Kuhn 1979a, viii). Yet despite the fact that the

Once Kuhn had taken the historical turn in philosophy of science, he came to believe a key issue in philosophy of science was to develop an understanding of what *really* settles disputes in science (see Kuhn 1992/2000, 108). His study of the history of science suggested that data do not play the role he and many others originally thought. Consequently, Kuhn felt that philosophers should aim to determine what it is that really secures consensus in a research community. Historical studies of science, he thought, could shed light on this issue.[6]

KUHN AND THE SOCIOLOGY OF SCIENCE

In the 1960s sociology of science was enjoying a period of significant growth. The Mertonian school had a near-monopoly.[7] The publication of *Structure*, though, inspired a new school in the sociology of science, the Strong Programme in the Sociology of Scientific Knowledge (see chapter 9 for more on this). Thus the historical turn that had affected philosophy of science so profoundly in the 1960s gave rise to equally unsettling developments in the sociology of science in the 1970s, as the Strong Programmers displaced the Mertonians.[8]

The Mertonians and the proponents of the Strong Programme differed profoundly. Whereas the Mertonians were sociologists by training, the early proponents of the Strong Programme were not initially trained as or by sociologists. Generally, they had training in the natural sciences and turned to a serious study of the history and sociology of science only in graduate school.[9] And whereas many of Merton's students were engaged in quantitative studies of science, the Strong Programmers conducted

---

view of science that Kuhn sought to replace was not an accurate representation of positivism, that is the view that came to represent positivism to many people (see Richardson 2007, 362).

[6] The claims of the historical school were ultimately subjected to empirical tests. Donovan *et al.* (1988) enlisted a number of philosophers to study a range of historical cases, to test "the theories of scientific change" developed by the historical school, specifically, the theories developed by Kuhn, Lakatos, Feyerabend, and Laudan (see Laudan *et al.* 1988, 6).

[7] On the influence of Merton and the rapid growth of sociology of science in the 1960s and early 1970s see Garfield (1980) and Cole and Zuckerman (1975). Incidentally, Merton (1977) attributes the development and growth of the sociology of science to the fact that science became "widely regarded as a social problem and as a powerful source of social problems" (111). Only then was there adequate interest in studying science sociologically. If this is in fact the case, it is not surprising that the Strong Programme studies of science would be perceived as anti-science. Some of these studies do, after all, draw attention to some of the social problems that science is implicated in.

[8] Just as some question whether Kuhn killed positivism (see Reisch 1991), one might question whether the Strong Programme killed the Mertonian school. There is some evidence to suggest that Merton's work in other areas of sociology besides the sociology of science, the sociology of deviance, for example, was also in decline in the early 1970s (Cole 1975, 200).

[9] It is worth noting that the early logical positivists "were [also] trained, in the first instance, as scientists, and spent their early professional lives in the environment of scientists" (Butts 2000, 198).

qualitative micro-studies of science, detailed studies of very specific episodes in the history of science, often focusing narrowly on one laboratory or research institution. Farley and Geison's (1974) study of the debate about spontaneous generation in nineteenth-century France and Steven Shapin's (1975) study of the debate about phrenology in nineteenth-century Edinburgh are typical. In the spirit of Kuhn's project, these studies sought to determine how consensus was *really* reached in science, that is, to discover how disputes in science were brought to a close.

Like Kuhn, the proponents of the Strong Programme maintained that observations are subject to multiple interpretations. Observation is thus incapable of unequivocally resolving disputes in science. But, unlike Kuhn, the proponents of the Strong Programme claimed that various contingencies that have *no epistemic import* determine which theory or hypothesis is ultimately accepted.

Although it is not one of the first-generation studies of the Strong Programme, Simon Schaffer's (1989) study of the reception of Newton's theory of color is a typical example of the work developed by these sociologists. Schaffer's study illustrates in a clear way the sort of things about the Strong Programme that concerned Kuhn and other philosophers of science. Schaffer suggests that Newton's theory of color won the assent of his peers only after he gained "control over the social institutions of experimental philosophy" (1989, 100). "After 1710 [Newton's] authority among London experimenters was overwhelming ... [allowing] carefully staged trials before chosen witnesses and the distribution of influential texts and instruments stamped with the imprimatur of collective assent" (100). Thus, on Schaffer's account, it was Newton's growing *power* that bridged the gap between data and theory, and secured the acceptance of his theory of color.

Incidentally, Schaffer's interpretation has been challenged. Alan Shapiro (1996), for example, has criticized Schaffer's account of the reception of Newton's theory of color on a number of grounds. Shapiro argues that, contrary to what Schaffer claims, "Newton's theory was ... established in Great Britain well before he had so much authority over 'London experimenters'" (132). Further, Shapiro argues that "Newton's 'control over the social institutions of experimental philosophy' in London [was] ... of minor consequence on the Continent" (132). On Shapiro's account, the "greatest attraction [of Newton's theory of color] was its comprehensiveness and explanatory power" (133).[10]

---

[10] I thank Trevor Pinch for drawing my attention to Shapiro's paper, and Peter Dear for drawing my attention to Schaffer's paper.

Despite the Strong Programme's interest in and enthusiasm for Kuhn's work, his reaction to their work was largely critical. He was unwilling to accept the skeptical implications of their sociological and historical studies of science. He described their project as "deconstruction gone mad" (Kuhn 1992/2000, 110; see also Kuhn 1991a/2000, 91). Specifically, he believed that by invoking negotiation, power, and interests to fill the gap between data and theory, the gap that methodology was unable to bridge, the Strong Programme left no role for nature to play in the process (see Kuhn 1992/2000, 109 and 110). Kuhn, though, insists that "you are not talking about anything worth calling science if you leave out the role of [nature]" (2000b, 317). He repeatedly notes that, contrary to what is suggested by the Strong Programme, the world is "not in the least respectful of observer's wishes and desires; quite capable of providing decisive evidence against invented hypotheses which fail to match its behavior" (Kuhn 1991a/2000, 101). Whether Kuhn's reading of the Strong Programme is a fair or accurate portrayal of their view is irrelevant for my purposes here. His reading of their view is similar to the standard way they are interpreted by philosophers (see, for example, Laudan 1984; and Friedman 2001).

It is not surprising that Kuhn reacted negatively to the Strong Programme. Indeed, the way Kuhn reacted is similar to the way many *philosophers* did and still do react to the Strong Programme. The proponents of the Strong Programme openly embrace a form of relativism that is thought to threaten the epistemic authority of science (see Pinch and Bijker 1984, 401).[11] Like most philosophers, Kuhn was not interested in threatening the epistemic authority of science and scientists. In fact, Kuhn was, if anything, an apologist for science (see Fuller 2000; Barnes 2003, 135). He took for granted the success of science.

It is worth mentioning that Kuhn was not opposed to the sociology of science in principle. In fact, in his 1969 "Postscript" to *Structure* Kuhn discusses the need for sociological studies of science as we seek to understand the nature of the "community structure of science," approvingly citing the work of a variety of sociologists, including Warren Hagstrom, Price and Beaver, Diana Crane, and Nicholas Mullins (Kuhn 1969/1996, 176n. 5). And in the Preface to *Essential Tension*, Kuhn describes his own

---

[11] Steve Shapin (1992) notes that Marxist external histories of science were "widely seen as … aggressive attempt[s] to devalue science" (339). Philosophers of science have tended to assume that the proponents of the Strong Programme had similar motives, thus leading many philosophers to regard "the rational" and "the social" as a legitimate contrast, where the latter threatens the former. This, though, is not Kuhn's concern with the Strong Programme.

work as "deeply sociological" (1977a, xx; see also Pinch 1979, 439).[12] But
he qualifies this claim about the sociological nature of his work with the
following remark: it is sociological but "not in a way that permits that
subject to be separated from epistemology" (1977a, xx).[13] Hence, Kuhn's
concerns with the studies of the Strong Programme were not motivated
by an animosity toward or an aversion to sociological studies of science
in general.[14] Nor were his concerns motivated by an animosity toward
externalist histories of science. In fact, Kuhn explicitly endorses externalist histories of science, regarding such studies as complementary to,
rather than competing with, internalist histories (see Kuhn 1968/1977,
119–20). Kuhn's reaction against the Strong Programme was due to the
general view of science that seemed to emerge from their studies, a view
that seemed to leave no role for nature in the process. Indeed, this same
complaint has been raised by Stephen Cole (1992) against constructionist
sociologists of science in general. As Cole explains, social constructionists
claim "that nature or 'truth' is irrelevant in determining what comes to
be accepted as scientific 'fact'" (Cole 1992, 136). Cole, it is worth noting,
was a former student of Merton's.[15]

Trevor Pinch (1982/1997) offers a different way to characterize this
concern that Kuhn raises about the research of the Strong Programme.
Speaking as one who is part of the group Cole refers to as constructionist sociologists, Pinch explains that "it makes little sense to break scientific activity down into its constituent social and cognitive parts" (473).
Indeed, this resistance to distinguishing between the social, on the one
hand, and the cognitive, on the other, is central to contemporary work in

[12] In his critical review of the second edition of *Structure*, Alan Musgrave (1971/1980) takes issue
with the sort of sociology of science that Kuhn appeals to in an effort to isolate scientific communities. Musgrave claims that such sociologists fail to take account of the scientific content of
the published articles they use to determine the membership of specific scientific research communities (40–41). It is interesting that this complaint was raised, given that philosophers seemed
even more enraged by the proponents of the Strong Programme, sociologists who consciously
sought to scrutinize the content of science.

[13] It is interesting to note that Reichenbach (1938/2006) begins *Experience and Prediction* by claiming that "epistemology [in its descriptive task] forms a part of sociology" (3).

[14] By the time Kuhn published *Black-Body Theory and the Quantum Discontinuity, 1894–1912* in
1978, many sociologists had come to believe that Kuhn had reneged on his earlier commitment to take the social dimensions of science seriously. Pinch (1979) explains that, given Kuhn's
account of the early days of this revolution in modern physics, it seems that he came to believe
that "the social dimension is only important in the dissemination of the new ideas; the physics
itself is immune to social factors" (439). Pinch thus thought Kuhn had changed his view since
the publication of *Structure*. To Pinch, this suggests that "the net result of Kuhn's [change of
view] is to maintain the division of labor whereby philosophers of science determine the criteria
of good science and sociologists study scientists' use of and deviations from such criteria" (440).

[15] I discuss Kuhn's relationship to constructionism in detail in chapter 9.

sociology of science. And it is, no doubt, part of the reason why philosophers misunderstand such work.

As the Strong Programme gained momentum, Kuhn began to rethink his philosophy of science. Although Kuhn believed that the Strong Programme's studies deepen our understanding of scientific controversies, he thought that "their net effect, at least from a philosophical perspective, [had] been to deepen rather than to eliminate the very difficulty they were intended to resolve" (Kuhn 1992/2000, 109). That is, there was still no clear understanding about what brings disputes in science to a close.

Kuhn does recognize that there are important similarities between his own view and the view of the Strong Programme. He claims that both he and the proponents of the Strong Programme believe that "facts are not prior to conclusions drawn from them, and those conclusions [or theories] cannot claim truth" (Kuhn 1992/2000, 115). But he believes that they part company when it comes to explaining what fills the gap between data and hypotheses. He objects to their "replacing evidence and reason by power and interest" (116).

KUHN'S NEW INSIGHT FROM HISTORY

Kuhn came to believe that he and the other philosophers who took the historical turn "overemphasized the empirical aspect of [their] enterprise" (Kuhn 1991a/2000, 95). Rather than treating the history of science as a body of data from which to construct a philosophy of science, he came to believe that the key insight that philosophers of science could gain from the history of science is a particular perspective on science (Kuhn 1991a/2000, 95). Kuhn calls it the historical perspective, or alternatively, "the developmental view" of science (95). This perspective, he claims, is the perspective that all historians bring to their subject (95). Kuhn believed that this perspective would give us greater insight into the nature of science and scientific knowledge. And it was in virtue of this perspective that he thought of his epistemology of science as an evolutionary epistemology.

Whereas philosophers traditionally conceived of scientific knowledge as a static body of belief, the historical perspective demands that we see science as "a process already underway" (1991a/2000, 95). As such, we must recognize that scientists are always working within a tradition, beginning their inquiries with a set of beliefs inherited from their predecessors. Kuhn claims that, given this perspective on science, we need to recognize that scientists are concerned with the evaluation of *changes* of belief rather than with the evaluations of *belief.*

Incidentally, even before the first generation of publications by the Strong Programme, Kuhn began to consider the significance of this change of perspective for the philosophy of science. In a paper presented in 1968 in which he discusses the relationship between the history of science and the philosophy of science Kuhn makes the following remark:

> The overwhelming majority of historical work is concerned with process, with development over time. In principle, development and change need not play a similar role in philosophy, but in practice … the philosopher's view of … science, and thus of such questions as theory structure and theory confirmation, would be fruitfully altered if they did. (Kuhn 1976/1977, 18)

Hence, in 1968 Kuhn had already begun to reflect on a developmental view of science. But he did not develop this line of thought further at that time. In fact, he did not spell out the implications of this shift until his presidential address to the Philosophy of Science Association in 1990 and his Rothschild Lecture the following year (see Kuhn 1991a/2000, 90–104; 1992/2000, 105–20). And because of his untimely death, the project was never completed.

Kuhn notes three important insights that follow from the historical perspective, insights that are contrary to the "positivist" view of science he rejected when he initially took the historical turn.

First, given the historical perspective, Kuhn claims that "the Archimedean platform outside of history … is gone beyond recall" (1992/2000, 115). Observations or data cannot and do not operate as a foundation upon which theories are constructed. Nor can they provide a theory-neutral basis from which to evaluate competing hypotheses (see Kuhn 1991a/2000, 95). Rather, scientists are always making evaluations of competing theories against the background of accepted beliefs, beliefs that may themselves come to be rejected in the future.

Second, Kuhn claims that, given the historical perspective, "comparative evaluation is all there is" (1992/2000, 115). That is, when confronted with a choice between a long-accepted theory and a new, alternative theory, scientists working within a tradition can evaluate the theories only comparatively. As a result, scientists are only ever in a position to conclude that one theory is *better* than the other.

Third, given the historical perspective, Kuhn claims that "no sense can be made of the notion of a reality as it has ordinarily functioned in philosophy of science" (1992/2000, 115). Kuhn explains that "within … the previous tradition in philosophy of science, beliefs were to be evaluated for their truth or for their probability of being true, where truth meant something like corresponding to the real, the mind-independent external world" (Kuhn 1992/2000, 114). Given that scientists can make only comparative judgments of the competing theories they have developed against a background of accepted beliefs inherited from their predecessors, the question of whether a particular theory mirrors reality seems misguided.

Kuhn claims that "seldom or never can one compare a newly proposed law or theory directly with reality" (1992/2000, 114). Instead, scientists must rely on what he calls the "secondary criteria": accuracy, scope, simplicity, and consistency (114). And because judgments of theories are comparative, scientists are judging whether a particular theory is simpler than competitor theories, or whether a particular theory is more accurate than competitor theories. But Kuhn wants us to see that there "is a price to be paid for" this shift in perspective. After all, as he explains, "a new body of belief could be *more* accurate, *more* consistent, broad*er* in its range of applicability, and also simpl*er* without for those reasons being any tru*er*" than the body of beliefs it replaces (115). Hence, it makes little sense to claim that a series of theory changes in a field are converging on the truth.

Given the historical perspective, "justification [aims] simply ... at improving the tools available for the job at hand" (1991a/2000, 96). It matters not whether our current background beliefs are true, for, even if they are not, our judgments about the relative worth of competing hypotheses can still be rational (see Kuhn 1991a/2000, 96; 1992/2000, 113).

The key to moving forward in developing a philosophy of science, Kuhn believes, is to abandon the traditional focus on truth and a mind-independent reality; however, Kuhn is not suggesting that "there is a reality which science fails to get at" (Kuhn 1992/2000, 115).[16] Kuhn is not that sort of skeptic. Rather, his point is that we can better understand the dynamics of scientific change and the nature of scientific knowledge if we see it as a process leading to increasing specialization. Kuhn thus claims that "what replaces the one big mind-independent world about which scientists were once said to discover the truth is the variety of niches within

---

[16] Kuhn's remarks on the role of truth in science are sometimes confusing (see Kuukkanen 2007). Insofar as the content of a theory can be expressed in propositions, Kuhn's view is similar to van Fraassen's (see van Fraassen 1980). Both believe that the truth values of theoretical claims are beyond our epistemic reach. To use David Papineau's (1996, 5) terms, they are both skeptical anti-realists rather than idealist anti-realists. But their skepticism is a selective skepticism, for both believe that science has been extremely successful at increasing our knowledge of the observable phenomena.

which the practitioners of the various specialties practice their trade"
(120).

Specialization is thus the means by which scientists improve their tools
for the job at hand. The success of science is achieved by the creation and
refinement of an ever-increasing number of theories and research communities. Indeed, as far as Kuhn is concerned, it is the proliferation of
specialties, not a continuous unrelenting march closer to the truth, which
accounts for the increasing accuracy that we see throughout the history
of modern science. By narrowing the scope of our investigations, as specialization inevitably does, we are able to develop conceptual tools that
better enable us to manipulate the world in predictable ways. This focus
on increasing specialization as the end of inquiry complements Kuhn's
(1992/1996, 170–73) insight that science is a process pushed from behind
rather than aiming at a goal set by nature in advance, a claim we will look
at in greater detail in chapter 6. And we will look at Kuhn's account of
specialty formation in more detail in chapter 7.[17]

Kuhn's proposal leads to a profoundly different understanding of the
goal of science from that which many philosophers have traditionally
assumed in their studies of science. Developing unifying theories has long
been regarded as one of the greatest achievements of science. And this goal
is intimately tied to the views that (1) scientific knowledge is cumulative,
and (2) successive theories in a field are converging on the truth. Kuhn's
emphasis on specialization reminds us that in the real world of science
scientists must often narrow the scope of their theories in their efforts
to realize their epistemic goals. Such a view is hard to reconcile with the
cumulative and convergent accounts of scientific knowledge. Thus, if we
take Kuhn's proposed shift in perspective seriously it will change our view
of the nature of science and scientific knowledge profoundly. Scientific
knowledge is in some sense a local knowledge.

I want to briefly address an objection that I anticipate to Kuhn's claim
that the increasing predictive accuracy is a consequent of specialization
rather than scientists developing theories that are closer to the truth
than the theories that preceded them. One might suggest that, contrary
to what Kuhn claims, specialization aids scientists in developing more
accurate theories *because* specialization is the means by which scientists
develop theories that are apt to be true. Hence, we need not explain the

---

[17] Even when Kuhn wrote the Postscript to the second edition of *Structure*, he already suggested
that specialization was the key to understanding why more recently developed theories are more
accurate than older theories (see Kuhn 1969/1996, 205–06).

increasing predictive success in terms of *either* specialization *or* scientists developing theories that are closer to the truth.

There are two problems with this line of reasoning. First, the theories scientists develop in various specialties are not consistent with each other. Indeed, new specialties are often created because the lexicon used to model one set of phenomena is unsuited to model another set of phenomena. This claim is discussed and defended in chapter 7, where I examine the creation of two new specialties in some detail, specifically, virology and endocrinology. Given the discrepancies between the lexicons employed in neighboring scientific specialties, we have little reason to believe that the various scientific lexicons can be unified into one true account of the world. Second, as Kuhn has noted, even in a single field that has undergone a series of changes in theory, there is little evidence that there is a convergence in ontologies. Kuhn gives the example of the transition from Aristotle's mechanics, to Newton's mechanics, to Einstein's general theory of relativity (see Kuhn 1969/1996, 206–07). Although Kuhn is confident that scientists have increased their knowledge of observables, he doubts that scientists have achieved a similar success with respect to knowledge of unobservable entities and processes.

In conclusion, the historical turn in the philosophy of science had a profound impact on the field. But philosophers slowly realized that it was far from clear what lessons they either should or legitimately could draw from the history of science. And with the publication of the historical studies of the Strong Programme philosophers developed an ambivalent attitude toward the history of science. The research of the Strong Programme rightly drew attention to the contingencies that affect scientists as they work in their labs and make sense of the data they have before them. But, like many philosophers, Kuhn was not satisfied with the view of science developed by the proponents of the Strong Programme. He was especially concerned about the skepticism that seemed to follow from their focused studies of the contingencies that affect scientists in their research, for example, power and interests. These contingencies, Kuhn believed, had no *epistemic* significance.

Kuhn never questioned the success of science. In his efforts to account for the success of science and to develop an epistemology of science Kuhn proposed a shift in perspective, one inspired by his study of history. The historical perspective recognizes that scientists always work within a tradition of accepted beliefs. And scientists advance their epistemic goals in specialist communities, communities that employ and develop concepts and instruments specially designed to advance their rather local goals.

Further, the evaluations scientists make concern changes of belief, which are comparative judgments. Hence, the traditional focus on truth in philosophy of science is misguided.

A final brief remark is in order about the following question: how are disputes in science resolved? Kuhn believes that when there is disagreement in a research community subjective factors ensure that the research efforts of the community are divided in such a way that the competing theories are developed. In time, as the theories are refined, their strengths and weaknesses are exposed. In this way, such disputes are resolved on the basis of a consideration of the epistemic merits of the competing theories. In chapter 9, there is a more complete and detailed discussion of this important issue.

CHAPTER 6

Truth and the end of scientific inquiry

Right from the beginning, that is, from the publication of *Structure*, Kuhn's epistemology of science was an evolutionary epistemology of sorts. What changed over time was the extent to which his epistemology was an evolutionary epistemology. Later in his life, the evolutionary dimensions of his epistemology were extended and developed. Scientific change, he came to believe, was even more like evolutionary change than he had initially thought.

Kuhn first compares scientific change to evolutionary change in the final pages of *Structure*. There, Kuhn (1962a/1996) challenges the common assumption that science is moving toward a fixed goal set by nature. Instead, he claims that science is like evolution, pushed from behind. Kuhn claims that this change in perspective, that is, seeing that science is not moving toward a goal fixed by nature in advance, is the key to understanding the nature and dynamics of scientific change.

I have two aims in this chapter. First, I aim to defend the claim that in some important sense science has no fixed goal. Thus, I aim to defend Kuhn's radical claim about the end of scientific inquiry. I argue that experimental findings do not constrain scientists' theorizing to the extent that many philosophers assume. Experimental results are not fixed, once and for all. Rather, the *significance* of experimental results is subject to change over time. Consequently, the goal of science is not aptly described as fixed in advance. Further, what data can and should be accounted for by a particular theory is something that is not determined by nature either in advance or once and for all. Indeed, we will see that scientific observation and measurement are also more complicated processes than philosophers traditionally assume.

Second, I aim to explain how Kuhn's evolutionary perspective on scientific change can enhance our understanding of the process. There are, as we will see, unsettling consequences that follow once we adopt Kuhn's evolutionary perspective. This is one reason philosophers of science have been reticent to adopt Kuhn's evolutionary perspective. Kuhn's evolutionary perspective has affinities with the perspective many contemporary sociologists of science adopt in their studies of science. But there are significant differences between Kuhn's view and the views of many sociologists of science. We saw this in the last chapter, and, as we will see further in chapter 9, as Kuhn developed his view he consciously sought to distinguish it from the views of sociologists of science, and especially the view of the proponents of the Strong Programme in the Sociology of Scientific Knowledge.

KUHN'S ACCOUNT OF THE END OF SCIENCE

In the final pages of *Structure*, Kuhn compares scientific change to evolutionary change and argues that, like evolutionary change, the development of science lacks a goal or *telos*. Instead, Kuhn sees the development of science as "a process of evolution *from* primitive beginnings – a process whose successive stages are characterized by an increasingly detailed and refined understanding of nature" (Kuhn 1962a/1996, 170; emphasis in original). Consequently, Kuhn claims, we must "relinquish the notion ... that changes of paradigm [that is, changes of theory] carry scientists ... closer to the truth" (Kuhn 1962a/1996, 170). In fact, he claims that the development of science is not "a process of evolution *toward* anything" (Kuhn 1962a/1996, 170–71; emphasis in original). In particular, he emphasizes that science does not draw "constantly nearer to some goal set by nature in advance" (Kuhn 1962a/1996, 171). Importantly, though, Kuhn does recognize that scientists are developing an increasingly detailed and refined understanding of nature (170).

Kuhn recognizes that he is presenting a view of scientific change that is contrary to the view commonly held by philosophers; however, he believes that the traditional account of the end of science, the appeal to truth, offers little insight into the nature of science and scientific change. He asks rhetorically: "does it really help to imagine that there is some one full, objective, true account of nature and that the proper measure of scientific achievement is the extent to which it brings us closer to that ultimate goal?" (Kuhn 1962a/1996, 171) Thus, Kuhn offers his evolutionary perspective as a means to developing a better understanding of the dynamics of scientific change. The traditional view, the truth-directed view, he suggests, is actually quite empty.

Kuhn anticipated that his proposed change of perspective would encounter resistance. In fact, he noted an interesting similarity between the reception of his account of scientific change and the reception of Darwin's theory of evolutionary change. Kuhn rightly noted that Darwin's theory of evolution by natural selection met with the greatest resistance on the issue of the elimination of teleology (Kuhn 1962a/1996, 171–72). Darwin's claim that *species have evolved*, on the other hand, encountered little resistance (in this regard, see Hull *et al.* 1978). Similarly, Kuhn expected that his account of scientific change would meet with resistance because of his suggestion that science has no *telos* (Kuhn 1962a/1996, 171).[1]

As Kuhn encountered criticism, he attempted to clarify his view of scientific inquiry. In fact, some have even suggested that as he clarified his account of scientific change, modifying it to address the concerns of his critics, he retracted the more interesting and controversial aspects of the account (see Fuller 2000, xii and 3). But Kuhn remained committed to this evolutionary analogy to the end (see Kuhn 2000). Indeed, as we saw in the previous chapter, Kuhn continued to invest more in developing his evolutionary perspective on science, identifying additional ways in which scientific change is like evolutionary change.[2]

Kuhn, though, did recognize that the evolutionary analogy can be taken too far. But even in *Structure* he notes two additional ways in which scientific change resembles evolutionary change: (1) in the biological world, it is the fittest competitor that survives; similarly, in a scientific dispute it is the fittest competitor that survives; and (2) just as biological evolution gives rise to a greater variety of species, scientific development leads to a greater variety of scientific specialties (Kuhn 1962a/1996, 172).[3]

---

[1] Hull *et al.* (1978) tested the hypothesis that older scientists are resistant to new theories by examining the reception of Darwin's theory of evolution in Britain. Interestingly, what Hull *et al.* (1978) tested was whether older scientists were less likely than younger scientists to accept the claim that *species evolve*, not Darwin's claim about the mechanism responsible for evolution, *natural selection*. Rightly, as Kuhn notes, natural selection, which denies that evolution has a *telos*, was not so readily accepted, even by those scientists prepared to accept that species evolve.

[2] Incidentally, as George Reisch (1991) has noted, Rudolf Carnap was intrigued by this dimension of Kuhn's account of theory change when he read about it in Kuhn's manuscript for the *International Encyclopedia of Unified Science*, that is, in the manuscript of *Structure*. In his capacity as associate editor for the *Encyclopedia*, Carnap explains in a letter to Kuhn that "you emphasize that the development of theories is not directed toward the perfect true theory, but is a process of improvement of an instrument. In my own work on inductive logic in recent years I have come to a similar idea" (reprinted in Reisch 1991, 267).

[3] Comparisons between evolutionary change and scientific change are common. Variations of (1), above, are developed by Ernst Mach (1896/1986, chapter 25); Karl Popper (1963, 1975/1981); Stephen Toulmin (1972); Bas C. van Fraassen (1980); David Hull (1988); and Ronald Giere (1999). It is difficult to determine the extent to which scientific change is like evolutionary change, and there is even some disagreement about Kuhn's view on the similarities between the two processes (compare Renzi 2009 with Reydon and Hoyningen-Huene 2010).

TWO EVOLUTIONARY MODELS CONSIDERED

Alexander Bird (2000) takes issue with Kuhn's claim that science does not have a fixed goal. Bird is motivated by the desire to defend the traditional view that successive changes in theory are aptly described as bringing us closer to the truth. As far as Bird is concerned, the success of our current theories is explained by the fact that we are getting increasingly closer to the truth. Bird is not alone in taking issue with Kuhn's biological metaphors (see Weinberg 1998; Renzi 2009; see also Reydon and Hoyningen-Huene 2010 for a critical discussion of Renzi 2009).

Bird does grant that scientific change is like evolutionary change in some respects. Specifically, he grants that "the proliferation of scientific fields bears a resemblance to speciation ... [and] Kuhn's picture of normal science interrupted by revolutions might be compared to the model of punctuated equilibria" (Bird 2000, 212). But he argues that the evolutionary analogy does not serve Kuhn's purpose of "denying that theories get closer to the truth" (Bird 2000, 212). Bird thus defends the traditional view of the end of inquiry, the view that has informed much philosophy of science.

In his effort to defend the traditional view, Bird compares two models of evolutionary change. The first model, the two-species model, involves two species evolving in competition and in reaction to each other. The second model, the one-species model, involves just one species evolving in a *relatively* fixed or stable environment. Bird argues that although the two-species model would support Kuhn's claim that science does not move toward a fixed goal, scientific change is more like the process of change represented by the one-species model. Consequently, Bird argues that insofar as scientific change is like evolutionary change, the process is similar to the process of evolution in the one-species model, a model that supports the traditional view.

Consider the differences between the two models. The two-evolving-species model involves a predator, like the cheetah, and one of its principal food sources, the gazelle (Bird 2000, 211–13). Natural selection is responsible for the fact that, on average, contemporary cheetahs run faster than earlier cheetahs. In the past, slower cheetahs had a more difficult time getting sufficient food than faster cheetahs with the result that faster cheetahs tended to live longer, and consequently tended to produce more offspring. Because cheetahs have passed this advantage on to their offspring, we encounter a much faster cheetah today than our ancestors encountered many generations earlier. But as Bird notes, the gazelle is also subject to selection pressures. Consequently, over the course of many generations, gazelles also run increasingly faster. Indeed, one of the reasons that contemporary cheetahs run as fast as they do is that gazelles have been increasing in speed. Similarly, one of the reasons that gazelles run as fast as they do is that cheetahs have been increasing in speed. And because cheetahs tend to catch slower gazelles, they have inadvertently contributed to the fact that contemporary gazelles now run faster than their predecessors.

The one-evolving-species model involves an animal like the giraffe evolving in a *relatively* fixed environment. According to Bird, natural selection is responsible for the fact that giraffes are taller now than they were many generations earlier. Taller giraffes have survival advantages over shorter giraffes other things being equal, for taller giraffes are able to access a greater range of edible leaves, many of which are unreachable to both shorter animals of other species and shorter giraffes. Because of their access to these food sources, taller giraffes tend to live longer than shorter giraffes, and consequently produce more offspring, with the result that the members of the species are on average taller now than they were many generations earlier. Optimally, the giraffe will reach a height suited to the height of the trees in its environment, which we have assumed is relatively stable. At that point, there will no longer be pressures or inducements for the giraffe to grow taller.

These two models of evolutionary change differ significantly in one important respect. Bird explains that "in the one-species scenario the test remains constant over time ... but in the two-species scenario the difficulty of the tests changes" (Bird 2000, 212). That is, whereas the *test* the giraffe encounters remains the same, the *tests* the cheetah encounters are increasingly difficult because "the gazelles are getting faster over time" (212). Indeed, as Bird notes, the gazelles are not only getting faster, they "are getting faster *because* the cheetahs are improving in speed" (212; emphasis in original). The cheetah's previous success *caused* the gazelles to get faster. Thus, whereas the giraffe has a fixed goal toward which it evolves, the cheetah is evolving toward a moving target.

Bird argues that the one-species model provides a "more accurate analogy for scientific development" (Bird 2000, 212). He claims that "the results of experimental tests do not change. A good experiment is one that is replicable; it gives the same results whenever performed" (212). Elaborating, Bird claims that "experimental tests do not change in a way that makes it more difficult for a theory to pass them, and even less do they do so *because* the theory is developing" (213). Hence, Bird concludes that the one-species model is superior because it "captures the idea that in science our theories may change but the features of the world that they respond to are what they are *independently* of our theories, and are by and large constant over time" (213; emphasis added). Further, Bird believes that just as the giraffe can "reach an optimal 'fit' with its (fixed) environment ... a theory can reach a [*sic*] optimal fit with the world, and this would be a true representation of it, since only true theories cannot be falsified" (213).

In summary, Bird objects to Kuhn's claim that science does not have a fixed goal for two reasons. First, Bird believes that the results of experimental tests in science are stable and unchanging. This claim concerns the extent to which observations are theory-laden. As far as Bird is concerned, Kuhn has exaggerated the theory-ladenness of observations. Second, Bird believes that the features of the world scientists seek to model are stable and more or less unchanging. They are not affected by our theorizing, because they are mind-independent. This latter claim I will refer to as the "independence claim." Bird insists that the world is more ready-made than Kuhn suggests.

Given that scientists have a fixed goal in both of these senses, Bird believes that successive theories in the history of a particular scientific field are aptly described as getting closer to the truth. He thus defends convergent realism. Bird thus advises us to resist the shift in perspective that Kuhn recommends.

EXPERIMENTAL RESULTS AND OBSERVATION RECONSIDERED

In the remainder of this chapter I defend Kuhn's claim that scientific change is better understood as a process moving from primitive beginnings rather than as a process moving toward a goal set by nature in advance. First, though, it must be shown that Bird is mistaken in claiming that science has a fixed goal. In this section my aim is to show that experimental results are less stable than Bird claims, and thus incapable of providing the sort of fixed goal that Bird seeks. Then I want to briefly examine observation and measurement and their role in science. I aim to show that both observation and measurement are far more complex processes than Bird implies.

One of the principal reasons that Bird believes that scientists have a goal set by nature in advance is that the results of experiments are so stable. It is in this respect that scientists are alleged to be more like giraffes growing to an optimal height which is determined in advance by their fixed environment, rather than like cheetahs that are continuously under pressure to change because their target, the gazelle, is changing. I aim to show that Bird exaggerates the stability of experimental results.

Bird regards the stability of experimental findings as evidence that scientists have a fixed goal. But, contrary to what Bird suggests, the stability of experimental results is not sufficient to fix the goal of scientists. After all, even though the *results* of a particular experiment may be stable and unchanging, the *significance of the results*, the bearing the results have on a dispute in science, are not fixed and unchanging. Hence, the same experimental results can, at one time, threaten to refute a hypothesis, and yet, at another time, be deemed irrelevant to either the confirmation or the refutation of a hypothesis. Consequently, contrary to what Bird implies, the results of experimental tests do not provide unequivocal constraints on scientists' theorizing. Hence, they do not fix the goal of science once and for all.

Disputes in science sometimes involve disagreements about what counts as a significant experimental result. Thus, a scientist can accept that a particular experimental test yields a specific result and yet still legitimately deny that the result is significant. This type of situation occurred in the debate about buoyancy between Galileo and his adversaries who sought to defend Aristotle's theory of buoyancy (see Drake 1970; Biagioli 1993). Galileo maintained that "the diversity of shapes given to this or that solid cannot in any way be the cause of its absolute sinking or floating" (Galilei 1612/2008, 85). That is, the shape of a body does not affect its buoyancy. Rather, according to Galileo, "all that shape influenced was the *speed* at which the body would sink or surface in the medium" (Biagioli 1993, 171; emphasis added; see also Galilei 1960, 33). But one of Galileo's adversaries, Ludovico delle Colombe, produced

a powerful experiment which seemed to refute Galileo's views on buoyancy … [He] showed that a sphere of ebony (a material with a specific weight superior to that of water) placed in water would sink, while a thin piece of the same material [weighing the same as the sphere] would remain afloat. From this he concluded that … buoyancy … depended upon shape. (Biagioli 1993, 171–72; see also Galilei 1960, 28)

Galileo's response to this experiment is revealing. He did not deny *the results* of his adversary's experiment (Galilei 1960, 26–45). Rather, Galileo denied the *significance* of the observation (Biagioli 1993, 173). He did this by qualifying or clarifying the intended scope of his theory of buoyancy. Galileo restricted the scope of his theory in a manner that rendered his adversary's experimental results irrelevant to his hypothesis about the relation between the shape of an object and its buoyancy.

First, he insisted upon a distinction between the behavior of bodies *in water* and the behavior of bodies *on the water's surface*. Galileo then claimed that his theory of buoyancy was intended to apply to bodies *in* a medium only, not to bodies *on* the surface of water (Galilei 1612/1960, 32; see also Biagioli 1993, 173). According to Galileo:

[T]o be in the water means to be placed in the water; and by Aristotle's own definition of place, to be placed implies to be surrounded by the surface of the ambient body; therefore the two shapes shall be in the water when the surface of the water shall embrace and surround them. But when my adversaries show the board of ebony not descending to the bottom, they put it not into the water but upon the water; there ... it is surrounded part by water and part by air. (Galilei 1612/2008, 92)

This distinction between being in water and on the water's surface enabled Galileo to set aside his adversary's alleged refutation of his hypothesis. And he was able to do so in a principled way. He insisted that the body resting on the water's surface was in fact in two mediums, the water and the surrounding air. Galileo thus rendered his adversary's experimental result irrelevant even though the result remained unchanged. Thin ebony boards placed on water do in fact float, as delle Colombe demonstrated.

The ebony experiment seems to have had a profound impact on Galileo's understanding of buoyancy. It was only after the ebony test was brought to his attention that Galileo was able to develop one of his more compelling arguments for his hypothesis that the shape of an object immersed in water does not affect whether it will sink or float. The argument is a *reductio ad absurdum* of his opponents' hypothesis that the shape of an object determines whether it floats or sinks in water. Galileo explains that:

[T]he plate of ebony and the ball, put into the water, both sink, but the ball more swiftly and the plate more slowly, and slower and slower according as it is broader and thinner; and the true cause of this slowness is the breadth of the shape. But these plates that descend slowly are the same that float when put lightly upon the water. Therefore, if what my adversaries affirm were true, the same identical shape in the same identical water would cause sometimes rest and other times slowness of motion. This is impossible ... Therefore, it must be something else, and not the shape, that keeps the plate of ebony above the water; the only effect of the shape is the retardation of the motion. (Galilei 1612/2008, 92–93)

This example shows that Bird is mistaken in claiming that experimental results provide a fixed goal for scientists, in the same way that the height of a tree provides a fixed goal for the giraffe. Granted, the results of the ebony experiment do not change. Thin ebony boards placed on the water's surface continue to float and ebony spheres continue to sink. In this sense, the phenomena are aptly described as fixed.

But the *significance* of the experimental results is not stable and unchanging. Hence, the implications of a thin floating ebony board placed on the surface of the water is indeterminate to some degree. If Galileo's hypothesis is understood to apply to both things *in* water and things *on* the surface of water, then the experimental result will refute his hypothesis. If, however, his hypothesis is merely about things *in* water then the experimental result is irrelevant as it concerns bodies *on* water. Thus, if the scope of his hypothesis is narrowed, then the experimental result is irrelevant. The significance of a particular experimental result thus varies depending upon how a hypothesis or theory is understood. And modifying a theory or hypothesis can change the significance of results.

It is important to recognize that Galileo's response to his adversary's experiment was neither deviant nor unscientific. Altering one's theory in light of new experimental results is a normal part of doing science.

Given the dynamic nature of experimental results illustrated here, it seems that there is a sense in which scientists have a moving target, contrary to what Bird would have us believe. One cannot know once and for all which experimental results need to be accounted for by any viable theory. Rather, as one develops one's theory, one may redefine which results matter.

The point here is not merely that observation is theory-laden. Even when observations remain fixed and can be described in a theory-neutral way, the significance that particular observations have in a dispute will depend upon what scientists purport to explain with their theories. Galileo *may* not have given much consideration to the distinction between bodies in water and bodies on the surface of water until his adversary produced the ebony experiment. He may even have assumed that his theory could apply to both sets of phenomena. And, once confronted with the experiment, he was not compelled by either logic or nature to subsume the phenomenon under the same theory that explains the behavior of bodies in water. He *could* legitimately narrow the scope of his theory of buoyancy and regard the anomaly as a phenomenon to be explained by a different theory. Determining the proper scope of a scientific theory is a difficult matter, one that scientists must sometimes face in disputes with their adversaries.[4]

---

[4] In chapter 7 we see how the discovery of hormones in the early 1900s affected physiologists' assumptions about the scope of their models and theories.

Nor is my point Quine's point that a hypothesis can be saved come what may. Indeed, Galileo did *save* his hypothesis by restricting its scope to bodies in water. But earnest scientists can be genuinely uncertain about what is entailed by their hypotheses. And confronting challenges can assist a scientist in clarifying what she means to claim. Too often philosophers assume that the implications of a theory are transparent to the scientists working with it. Philosophers too often assume that scientific theories can be expressed in axioms and their contents can be deduced from the axioms. Such an assumption, though, is at odds with the real world of science, where theories are developed and refined in response to new data and criticism.

Recall Kuhn's theory of scientific discovery, briefly discussed in chapter 3, above (1962b/1977). According to Kuhn, scientific discoveries are not psychological events in the minds of scientists involving the recognition of some aspect of the world that had hitherto eluded our attention. Rather, discoveries are complex, convoluted affairs that unfold over time. They are met with resistance, and they have to be worked out. It is often only after the fact that an event can be described as *the* discovery. And sometimes, as in the case of the discovery of oxygen, it can remain unclear exactly when the discovery occurred. In such cases, often the best we can do is to identify a window of time outside of which it is clear that the discovery did not occur.

Kuhn is not alone, even among philosophers, in noting the instability of the significance of data. Helen Longino (1990) makes a similar point, arguing that "in the absence of [background] beliefs no state of affairs will be taken as evidence of any other" (44). Because scientists working on the same research topic or on the same issue often share the same background assumptions, they often do not take note of the role played by background assumptions in stabilizing the significance of data. But two scientists working from different background assumptions may be led to different conclusions about the significance of a particular experiment. Moreover, each can justify her own interpretation, provided each is granted the background assumptions she has brought to her inquiry. Bird's argument for the stability of experimental results seems to have no regard for the fact that the significance of such results is indeterminable when separated from background assumptions. Background assumptions play an indispensable role in scientific reasoning.

It is worth emphasizing that it is not just experimental results that are malleable. Even in scientific fields where experiments are uncommon, scientists must learn to make judgments about how to classify the phenomena. And these judgments are always subject to revisions. The world does not come ready-made or packaged into clearly defined kinds. Hence, observation itself is negotiated to some extent.

In a recent article, Lorraine Daston (2008) discusses the challenges scientists encountered as they tried to develop a uniform set of classifications for describing clouds. The efforts of nineteenth-century meteorologists culminated in the production of *The International Cloud Atlas*, which "was meant to make clear-cut scientific objects out of evanescent, protean clouds by teaching observers all over the world ... to see things in unison" (104). Observation in science is not a straightforward affair. Rather, as Daston explains:

[S]cientific perception – especially when elevated to the level of systematic observation, often in carefully designed setups – is disciplined in every sense of the word: instilled by education and practice, checked and cross-checked both by other observers and with other instruments, communicated in forms – texts, images, tables – designed by and for a scientific collective over decades and sometimes centuries. (2008, 102)

Thus, the world of science is far less stable than Bird claims. Even observations, at least the sorts of observations that scientists work with, are more malleable than Bird has led us to believe.

Kuhn also suggests that even *measurement* is a more complex process than is generally recognized. He argues that sometimes measurement yields results that support a theory only when scientists assume the theory in question. For example, Kuhn claims that chemists only learned "how to perform quantitative analyses that displayed multiple proportions," the sort of results needed to support Dalton's law of multiple proportions, when they let "Dalton's theory lead them" (1961/1977, 196). Philosophers generally assume that measurement is a straightforward exercise, and thus capable of functioning as an impartial arbiter between competing theories. Kuhn gives us reason to challenge this assumption. Measurement is a complex activity, sustained by a research community's norms and standards, and shaped by the expectations one brings to the exercise.

Thus, we can see that the experimental results, observations, and measurements that scientists seek to account for with their theories are not unchanging and stable. Experimental results, observations, and measurements, once collected, do not provide unequivocal constraints on the theories scientists develop. Granting this, though, does not pose as great a threat to science as Bird seems to think.

I think that some of Bird's concerns with Kuhn's view are based on a mistaken understanding of Kuhn's claim that the goal of science is not fixed. Contrary to what Bird suggests, Kuhn does not deny the "independence claim," the claim that the *features* of the world which our theories respond to are what they are independent of our theories. Some clarification is in order here. Kuhn does not believe that the Earth changed its behavior or position in the solar system when early modern astronomers changed theories. Rather, early modern astronomers saw the errors of their predecessors. Similarly, Kuhn does not believe that tectonic plates were created in the mid 1960s with the acceptance of the theory describing their effects on the Earth's surface. Thus, even though Kuhn believes that the significance of experimental results is subject to change as theories change, he does not believe that the features of the world that our theories aim to model are subject to change as we change theories. Kuhn insists that the mind-independent world constrains what scientists can say about it (see Hoyningen-Huene 1989/1993, 33–34). What Kuhn does claim, though, is that when we change theories we often attend to different mind-independent features of the world. Thus, before and after a revolution in science, the scientists working in a particular field do not attend to the same features of the world. We will examine this dimension of Kuhn's view in greater detail in chapter 9.

In summary, the considerations raised above suggest that Bird's two-species model provides a more accurate model of the process of scientific change than his one-species model. But I believe that neither model captures Kuhn's view. Both models assume that there is *some* target toward which science is aiming. In the one-species model the giraffe aims for the optimal height, a target fixed by the height of the trees, and in the two-species model the cheetah aims to be faster than the gazelle, a target that moves. In developing the evolutionary analogy Kuhn's point is to show that the process of scientific change is best understood as a process that does not move toward *any* goal at all. Rather, Kuhn insists that, like the process of evolution by natural selection, scientific change is a process that is best conceptualized as driven from behind.[5]

---

[5] There are additional problems with Bird's two models. First, the one-species model is in fact a two-species model, for the trees that function as the goal of the giraffes are a biological kind, and thus subject to change. I do, though, understand Bird's point. He appeals to the one-species model as an idealization. Second, the two-species model is far too simplistic insofar as there is only one changing variable, speed. This gives us the false impression that there is a single direction to the changes in biological evolution. Bird would certainly like us to think that there is a single direction in which the changes in science are headed. But Kuhn's intention is to challenge this assumption. In chapter 9, we see that there is reason to follow Kuhn. New theories often

THE KUHNIAN PERSPECTIVE ON SCIENCE

In this section, I want to examine what can be gained from adopting Kuhn's perspective on scientific change, that is, the perspective that sees scientific change as a process *from* primitive beginnings rather than as a process *toward* a goal set by nature in advance. Further, I want to re-examine the popular assumption that motivates Bird's criticism of Kuhn, the assumption that later theories are generally closer to the truth than earlier theories.

Let us consider what can be gained from seeing scientific change from an evolutionary perspective, as Kuhn suggests. Rather than aiming for a goal set by nature in advance, Kuhn wants us to see that scientists' research agendas are set by their predecessors and peers. The problems that scientists are expected to address are defined by the work of their predecessors and peers (see also Toulmin 1972, 148). The concepts, instruments, and models with which they approach these problems are supplied by their predecessors and peers. And the data that scientists are expected to account for when developing their theories are determined by the previous developments in the field. Even the justification of or warrant for a change in theory is grounded in the past. It is because a new theory can resolve an anomaly that arose out of the previous normal scientific research that it comes to replace its predecessor.[6] In this respect, science is very backward looking. Caution is in order here. The term "backward looking" is not intended to imply something pejorative. Instead, the term is meant to acknowledge the tradition-bound nature of science, a theme explored in the previous chapter and further in the final part of the book.

The significance that past standards, concepts, and accomplishments have for later developments in science can be seen in any episode in the history of science. For example, when Newton was developing his physical theory he was not concerned with many of the goals of the physicists who followed him. He did not, for example, consider particles moving at the speed of light. Newton did not even conceive of such phenomena. Rather his concerns were continuous with the concerns of his

---

introduce changes in what scientists seek to model. In this way, the goals in a scientific field can be subject to quite complex changes that are not aptly described as heading in a single direction.

[6] Stephen Toulmin makes a similar point. He claims that "later models and concepts owe their legitimacy to having resolved problems for which earlier models and concepts were inadequate" (Toulmin 1972, 149).

predecessors: understanding the behavior of falling bodies, pendulums, motion propagated through fluids, projectiles, and orbiting planets and satellites (see Newton 1726/1999).

This is as we should expect, for, as Kuhn explains, "the extent of the innovation that any individual [scientist] can produce is necessarily limited, for each individual must employ in his research the tools that he acquires from a traditional education, and he cannot in his own lifetime replace them all" (1957, 183). Hence, changes in science are best understood as responses to existing problems, not as attempts to get at a description of the world as it *really* is. Kuhn's recommended shift in perspective thus draws our attention to the significance that past standards, concepts, and accomplishments play in later developments in science. It reminds us that scientific research is tradition-bound.

This is a radical perspective, at least among philosophers. Philosophers of science tend to assume that experimentation, observation, and measurement are capable of leading to unequivocal judgments about competing hypotheses. Sociologists of science, though, have seen that scientific inquiry is not so straightforward. In fact, Kuhn's evolutionary perspective has affinities with the views of many contemporary sociologists of science, including Barnes and Bloor, Steven Shapin, Harry Collins, and Bruno Latour. Like Kuhn, these sociologists of science recognize that theory choice is often underdetermined by the available data. Sociologists also generally do not assume that the accepted theories are accepted because they are true.

Latour's (1987) contrast between science in-the-making and ready-made science achieves a similar aim. In ready-made science the results are widely accepted. When we examine ready-made science we tend to think that scientific discoveries are inevitable. In contrast, when we examine science in-the-making, where a dispute is still unresolved, we are reminded that it is difficult for scientists to distinguish between a genuine discovery and the "discovery" of something that will soon come to be rejected. The alleged discovery of n-rays, for example, was initially greeted with the same enthusiasm and interest as the discovery of x-rays, made a few years earlier. But, in time, it was recognized that n-rays do not exist. Indeed, Kuhn's theory of *discovery* reminds us that scientific discoveries are not aptly characterized as the mere finding of something that was already there waiting to be discovered. Discovery in science is a complex process both constrained by and made possible by the accepted conceptual resources. Conceptual, mathematical, and instrumental resources often need to be developed and refined in order to complete the discovery process.[7] We will consider Kuhn's view on the inevitability of scientific discoveries in more depth in chapter 9.

It seems that the change in perspective that Kuhn recommends is also better suited to advancing our understanding of scientific disputes that occurred in the past. Rather than inadvertently projecting contemporary standards and methods on to earlier scientists as we seek to understand an episode in the history of science, Kuhn's perspective reminds us to be mindful of factors that were relevant to the scientists involved, factors that no longer seem relevant to us given our interests and the interests of contemporary scientists and philosophers. For example, McMullin (1984) notes that in the past, even in the seventeenth century, metaphysical and theological factors functioned as epistemic factors in scientific disputes. That is, a scientist and his or her audience would take such factors "to be a proper part of the argument he or she is making" (McMullin 1984, 129). If we are going to understand the dynamics of scientific change then we will need to examine past episodes from the perspective of those involved. Kuhn's evolutionary perspective puts us in a better position to understand the scientific changes of the past.

In contrast, it is unclear what insight one stands to gain from the traditional view, the view that scientific change is moving toward a goal set by nature in advance. The traditional view seems to gain whatever credibility it has illegitimately, as a result of the fact that traditional views are often accepted uncritically. In the history of astronomy, earth-centered models of the solar system held early modern astronomers back for ages. Similarly, truth-centered models of scientific change are holding philosophers of science back.

Given Kuhn's perspective on scientific change, successive theories are not aptly described as getting closer to the truth. From the Kuhnian perspective, successive changes in theory are indeed *improvements*, as one would expect. But they are improvements of a local and relative nature. A new theory is better than the one it replaces.[8] But, given the comparative nature of theory evaluation, it is questionable to draw the inference that the superior of two competing theories is *true*. Given the Kuhnian perspective, there is no basis for making such a claim. Further, because scientists do not have a target set by nature in advance, there is no standard against which we could measure scientists' success through a series

---

[7] Influenced by Kuhn, Tom Nickles has developed a similar view of the discovery process. According to Nickles' multi-pass conception of scientific inquiry the process of inquiry is "slow and highly serial, articulated, [and] segmented" (1997, 19).

[8] Todd Grantham (1994) reaches a similar conclusion about scientific progress.

of changes in theory. As research agendas, conceptual schemes, scientific instruments, and the significance of experimental findings change, measures of success change. Hence, it is futile to claim that after a series of changes in theory scientists are closer to the truth.

Even given Larry Laudan's (1984) reticulated model of scientific change, which was explicitly designed to improve on a popular reading of Kuhn's view, new theories can be evaluated only relative to the theories they replace. According to Laudan, the reticulated model "forces on us the recognition (which should have been clear all along) that progress is always 'progress relative to some set of aims'" (66). And as long as our aims are changing we are limited in the scope of the claims we can make about progress through changes of theory.

My aim in this chapter has been to identify a radical change in perspective that Kuhn recommends for philosophers of science, a change that, though introduced in *Structure*, was largely ignored by philosophers. According to Kuhn, science does not have an end set by nature in advance. Our goal is not to develop an accurate reflection of the world as it is independent of our theorizing. Rather, the goals of scientists are local, subject to change, and determined, to a large extent, by the goals, practices, and accomplishments of their predecessors. In this respect, there are interesting similarities between Kuhn's view and the views of many contemporary sociologists of science.

But despite the affinities between Kuhn's perspective on science and the perspective of many contemporary sociologists of science, there are also significant differences between their views. Although I discussed some of these differences briefly in the previous chapter, we consider the issue in greater detail in chapter 9. In particular, we will see that there are striking differences between the type of social constructionism that many sociologists of science endorse and Kuhn's constructionism.

CHAPTER 7

Scientific specialization

One of the most striking forms of scientific change is the rapid and seemingly endless growth of new scientific specialties. Nicholas Rescher (1978) notes, for example, that the number of specialties in physics has grown from 19 in 1911, to 100 in 1954, and reaches 205 in 1970 (229, table 3). Philosophers of science seldom discuss this dimension of scientific change. Specialization has been neglected by philosophers, in part, because they have tended to emphasize the value of unification in science (see, for example, Friedman 1974; and Kitcher 1993). Unifying theories deepen our understanding of the natural world by revealing connections between otherwise disparate phenomena. Specialization seems antithetical to this goal. Specialization can and often does create barriers between scientists. Thus, for many philosophers, specialization is seen as either an impediment to developing unifying theories or a temporary resting state along the way to developing unifying theories.

In chapter 1, in the brief discussion of the discovery of x-rays, we saw that theory replacement is not the only response that a research community has when it encounters persistent anomalies. Sometimes, as Kuhn notes, anomalies are dealt with by creating a new scientific specialty, a new research community that has as its concern the study of the previously anomalous phenomena. Further, the creation of new specialties is by no means a temporary state in the development of science. Rather, as Kuhn claims, specialization often serves to advance our epistemic goals in science. Hence, philosophers need to develop a better understanding of the role and effects of specialization in science.

In this chapter, I examine Kuhn's account of specialty formation. Kuhn's account is important for at least two reasons. First, Kuhn's account of specialization recognizes the important role played by epistemic factors in the creation of new specialties, factors such as conceptual developments. Sociological accounts of specialization tend to emphasize the influence of the sorts of factors that philosophers regard as external factors in the creation of new specialties. Consequently, Kuhn's account of specialization is an important correction to the sociological accounts. Second, as Kuhn developed his epistemology of science, specialization came to play a greater role. That is, Kuhn came to believe that specialization was a key means by which scientists realize their epistemic goals. Unfortunately, Kuhn never presented his views on scientific specialization in a systematic fashion. Nor did he make the topic the focus of a single paper. My aim is to bring together a coherent Kuhnian account of specialization.

SOCIOLOGICAL AND HISTORICAL STUDIES OF SCIENTIFIC SPECIALIZATION

In an effort to appreciate Kuhn's account of specialization, it will be useful to examine what sociologists and historians have said on the topic. Their accounts of specialization will aid us in seeing what is important and novel about Kuhn's account.

In stark contrast to philosophers of science, sociologists and historians have studied specialization with intense interest. In fact, Harriet Zuckerman (1988) notes that the sociological study of scientific specialization was itself, for some time, a sociological specialty (535). Interest in specialization, though, has since passed, as sociologists of science became more engaged in micro-studies of science, studies of individual laboratories or research facilities.

Sociological accounts of scientific specialization tend to focus on social and instrumental changes as the cause of the creation of new specialties. Although sociologists of science are correct in pointing out that the creation of a new scientific specialty often involves changes in the social organization of science, or changes in instrumentation, they have tended to contrast the social and instrumental with the conceptual, and given greater emphasis to the former at the cost of recognizing the role of the latter. Sociological studies of specialization thus tend to reinforce the traditional assumption of philosophers that specialization is of little significance to an epistemology of science.

In this section I examine a variety of sociological and historical studies of scientific specialization. Because the literature on specialization is so vast, a comprehensive survey is not possible. Instead, I begin by examining two very influential pioneering studies. I argue that both of these pioneering studies treat the development of a new specialty as essentially a change in the social organization of science, and thus treat any *conceptual changes* that one might associate with specialization as derivative.

Further, both of these studies imply that there is a single type of cause of the creation of new scientific specialties, though they differ about what that cause is. I then briefly review the sociological studies of specialization that followed these pioneering studies. These second-generation studies rightly challenge the assumption that the same type of cause is responsible for the creation of all scientific specialties. They are thus open to a plurality of causes. But I argue that, like the pioneering studies, these studies continue to treat the conceptual, epistemic, or cognitive changes associated with scientific specialties as secondary to the social and instrumental changes associated with the process.

Let us begin with the pioneering accounts. According to Joseph Ben-David and Randall Collins (1966/1991), the creation of a new scientific specialty is a consequence of scientists carving out a new professional niche in an effort to create a new social role (50). When an existing field shows little promise for career advancement, ambitious and able young scientists will seek means to create a new discipline, field, or specialty. Ben-David and Collins support their account with a case study of the creation of experimental psychology as a discipline in Germany in the late nineteenth century. As they note, before experimental psychology became a distinct discipline, "the subject matter of psychology was divided between speculative philosophy and physiology" (53). In the mid nineteenth century, the field of physiology underwent a period of rapid expansion (63). In a relatively short period of time, between 1850 and 1864, many of the university positions in physiology were filled by *young* men. As a result, there was little opportunity for career advancement for the next generation of physiologists (63). Positions were blocked by the recently appointed generation of men who still had full careers ahead of them. This led a number of promising young scientists trained in physiology to create new career opportunities for themselves by applying the methods of physiology to problems in psychology. In this way, they created a niche for themselves. In turn, they gave birth to the specialty of experimental psychology.

Ben-David and Collins suggest that this pattern of development that led to the creation of psychology as a specialty is typical of the process by which new specialties are created. Crowding in an existing field leads young scientists to develop a new specialty in an effort to secure rewarding employment. Importantly, Ben-David and Collins insist that conceptual developments in the study of the human mind were not responsible for the creation of experimental psychology as a discipline (50). In fact, Ben-David and Collins believe that, generally, "the ideas necessary for the creation of a new discipline are usually available over a relatively prolonged period of time and in several places" (50). Hence, ideas cannot be the driving force in creating new specialties. In fact, Ben-David and Collins claim that if ideas were sufficient for the creation of a new discipline, we should have expected psychology to develop as a discipline first in either France or Britain (67–69). But, instead, it was in Germany that the field of psychology was created.

Let us now consider the second pioneering account of specialization. Derek de Solla Price (1963/1986) suggests that the chief factor that leads to the creation of a new specialty is the demand to make effective research possible. Scientific research is done by humans, and our limited cognitive capacities place significant constraints on the organization of science. Because there is a limit to how much people can read, each scientist can attend to only a finite and rather small portion of the continuously growing body of scientific literature. As more and more people get involved in science, and more and more journals publish more and more articles, each new generation of scientists confronts a larger body of scientific literature. In fact, Price estimates that by the early 1960s there were already over 10,000,000 published scientific articles. And the number of publications was doubling every fifteen years. Price believes that the various subfields in science are a consequence of scientists carving out manageable bodies of literature.

Price sought to determine how large a body of literature could be such that a group of scientists could both produce it and keep abreast of developments in the literature. He suggests that 100 scientists each producing 100 articles in the course of their careers would produce a body of literature containing 10,000 articles (1963/1986, 65). Price believes that scientists could read that much over the course of their careers. Consequently, he concludes that a specialty must consist of about 100 publishing scientists. Such a community could keep abreast of the literature they produce (65).[1]

There are three features that these two pioneering accounts have in common. First, both accounts are premised on the assumption that conceptual developments in science are not what lead scientists to create new specialties. Price does not attribute any role to conceptual developments, and Ben-David and Collins explicitly argue that such developments are

---

[1] I have taken issue with Price's estimate of the size of scientific specialties elsewhere (see Wray 2010). Significantly, in making his calculation, Price failed to account for the fact that the average scientist publishes only 3.5 articles in her career, not 100 articles as he assumes in his calculations. Consequently, I have argued that specializations are probably both *larger* in terms of personnel than he suggests, perhaps involving between 250 and 600 scientists, but *smaller* in terms of research literature they produce and read, perhaps containing about 2,500 articles.

insufficient to cause scientists to create a new specialty. Hence, both of these pioneering studies privilege social change in the creation of a new specialty, and treat conceptual changes as derivative. So, insofar as specialization in science leads to conceptual innovations these are derivative of the social changes that created the new social group. Second, both accounts are premised on the assumption that crowding in an existing field leads to the creation of a new specialty. Price believes crowding has a different function from that hypothesized by Ben-David and Collins. Rather than driving young scientists in search of new career opportunities, Price believes that crowding leads communities of scientists to narrow their area of research in an effort to avoid being overwhelmed by the growing body of research. Only by narrowing their area of research, and thus creating a new specialty, are scientists able to effectively manage the continuously growing body of literature. Third, both accounts are mono-causal, premised on the assumption that there is one type of cause that leads to the creation of new scientific specialties.

We saw in chapter 5 that the 1970s marked a period of extensive growth in the sociology of science in general (see Cole and Zuckerman 1975, 146–48, 165). During that time the topic of scientific specialization attracted great interest among sociologists of science (Zuckerman 1988). Consequently, the literature on this topic is vast, but the most important of these studies were Nicholas Mullins' (1972) "The Development of a Scientific Specialty," David Edge and Michael Mulkay's (1976) *Astronomy Transformed*, Daryl Chubin's (1976) "The Conceptualization of Scientific Specialties," and an anthology edited by Lemaine *et al.* (1976b), *Perspectives on the Emergence of Scientific Disciplines*. Rather than provide a survey of second-generation studies, I want to draw attention to two ways these studies differed from the pioneering studies.

First, a number of the second-generation sociological studies examined the impact that new instruments and the development of instrumentation played in the creation of new specialties. For example, Edge and Mulkay (1976) provide a detailed account of the ways in which new instruments shaped the field of radio astronomy. Similarly, John Law (1976) argues that developments in instrumentation played a significant role in the creation of the field of x-ray crystallography. These studies thus suggest a need to recognize the range of causes that lead to the creation of new scientific specialties. But the authors of these studies tend to assume that developments in instrumentation are distinct from conceptual developments. This assumption is most evident in Law's (1976) study of x-ray protein crystallography. Law distinguishes between "'technique,' 'theory,' and 'subject matter' specialties" (123), implying that the techniques employed in x-ray crystallography could have been developed independent of the accompanying theoretical developments.

Second, a number of second-generation sociological studies acknowledge the complexity of the process by which a new specialty is created (Lemaine *et al.* 1976a). Indeed, these studies suggest that there might not be a single type of cause responsible for the creation of all scientific specialties. Further, the authors of these studies began to recognize that the development of a new specialty involves social changes, the focus of the pioneering studies, and conceptual or cognitive changes (Edge and Mulkay 1976, 364; Mulkay and Edge 1976, 153; Worboys 1976, 77; Lemaine *et al.* 1976a, 1). Some, like Chubin (1976), even question the derivative role that epistemic developments were assigned in the pioneering studies. Expressing concern, Chubin notes that sociological studies of scientific specialization tend to take "social structure as an antecedent of specialties," which "seemingly denies the possibility that intellectual events and the relations they engender give rise to a social structure that we treat as a specialty" (1976, 449). Becher and Trowler (2001) suggest that, though Chubin was concerned about the derivative role assigned to conceptual changes in the creation of new specialties, he was unable to explain how conceptual changes could be responsible for the creation of new specialties. Hence, these studies continued to treat epistemic developments as secondary.

Unfortunately, the interests and attention of sociologists of science shifted away from the study of scientific specialization before this concern was adequately addressed. Zuckerman (1988) notes that "with the shift in research attention to the microsociology of scientific knowledge, the study of specialties ... declined markedly amongst sociologists" (535). But in a review article on sociological studies of scientific specialization published in the late 1980s, Zuckerman (1988) notes that "the reasons that led sociologists of science to study the development of specialties in the first place still appear to be valid" (535). The study of specialization ended prematurely. Like Zuckerman, I am concerned that we have yet to develop an adequate understanding of the role that conceptual changes can and do play in the creation of new scientific specialties. In this respect, it seems that we do not understand the epistemic dimension of scientific specialization.

SCIENTIFIC SPECIALIZATION AND CONCEPTUAL CHANGE

In this section, I want to examine Kuhn's account of specialty formation. Kuhn discusses specialization in various essays collected together in

The Road since Structure. In his developed epistemology of science, special-ization plays a key role in scientific change, as discussed briefly in the pre-vious two chapters. It is unfortunate that he never wrote a paper devoted exclusively to this topic. As a result, there has been remarkably little uptake of or response to what he said on the issue. My aim is to synthe-size his remarks into a coherent account of specialization. Importantly, his account meets the challenge left unanswered by the sociological accounts of specialization. That is, Kuhn focuses on the *epistemic* dimension of the change. Consequently, his account should be of great interest to philoso-phers of science, as well as sociologists and historians of science.

According to Kuhn (1991a/2000), sometimes in their efforts to accom-modate or account for a persistent anomaly scientists are led to create a new scientific specialty (97). When scientists in a field encounter persistent anomalies and are unable to resolve the crisis with the resources provided by the prevailing theory, a new theory will inevitably be developed. The new theory is designed to resolve or accommodate the persistent anomal-ies. But sometimes the new theory will not be able to serve the purposes of all those working in the field. As a consequence, part of the field as it was conceived before the change becomes a new field or specialty. Thus, old theories are not necessarily discarded. Sometimes they come to be employed in a restricted domain. And a new theory is created to account for the persistent anomalies.

In chapter 1 we saw that as Kuhn developed his account of scientific change in response to criticism, he modified his view of scientific revolutions in subtle but important ways. These modifications have important implica-tions for understanding the process by which a new scientific specialty is created. Let us begin by briefly re-examining how he reconceived the notion of a scientific revolution. In Kuhn's developed account of scientific revolu-tions, the notion of a taxonomic or lexical change replaces the earlier prob-lematic notion of a paradigm change. Thus, a scientific revolution involves a taxonomic or lexical change. For example, with the Copernican revolution astronomers replaced a lexicon in which "planet" denotes a wandering star as opposed to a fixed star with a lexicon in which "planet" denotes a celes-tial body that orbits the Sun (Kuhn 1987/2000, 15).[2]

Kuhn believes that "the transition to a new lexical structure, to a revised set of kinds, permits the resolution of problems with which the previous

---

[2] Incidentally, the term "planet" was not used in a uniform way even by ancient astronomers. For example, as Dreyer (1963) notes, Ptolemy spoke of "the five wandering stars" "although it was more usual among the ancients to speak of seven planets," counting the Sun and Moon among them (196).

structure was unable to deal" (1993/2000, 250). For example, the new lexicon introduced during the Copernican revolution enabled astronomers to explain retrograde motion. The planets do not appear to go backward because they move on epicycles and deferent circles as they orbit the earth, as the Ptolemaic theory seems to imply. Rather, their apparent backward motion is a function of the fact that the Earth, the place from which we observe the other planets, is itself in motion around the Sun. When in the course of *our* orbit we pass another planet relative to the backdrop of the fixed stars, the other planet will *appear* to move backwards. In every scientific revolution, a new lexicon is introduced in order to resolve an outstanding problem.

According to Kuhn, sometimes the taxonomic changes that are introduced to accommodate a persistent anomaly affect only a sub-set of a scientific research community. And "what lies outside of [the new lexicon] becomes the domain of another scientific specialty, a specialty in which an evolving form of the old kind terms remain in use" (1993/2000, 250). Hence, a new specialty is created when a lexical change affects only part of a research community, and the old lexicon can still be employed *effectively* in a restricted domain.

But not every anomaly can be accommodated in such a way as to permit the continued use of the earlier taxonomy in a restricted domain. For example, Copernicus' proposed taxonomic change could not be adopted without replacing the Ptolemaic lexicon. In cases such as this one, there is no need to create a new specialty, for the whole of the existing research community adopts the proposed change.[3]

Hence, in mature fields, Kuhn claims, the history of science consists of periods of normal science punctuated by either (1) episodes of theory change, that is, scientific revolutions, or (2) episodes of specialty formation, where a new field branches off from a parent field. The result, in the long run, is a proliferation of fields, each concerned with a rather narrow domain. The gains in accuracy that scientists have achieved over time are a consequence of these two processes.

Given Kuhn's account, specialization involves both epistemic and social changes. On the social dimension, scientists who once had

---

[3] Mulkay (1975) describes Kuhn's paradigm-related account of scientific development as a closed model, for it stresses "the existence of scientific orthodoxies" (512). Mulkay believes that such accounts are unacceptable for they "make scientific innovation highly problematic" (513). Instead, Mulkay recommends a branching model of scientific development, according to which "new problem areas are regularly created and associated social networks formed" (520). Given Kuhn's account of specialization presented here, Kuhn's view is aptly described as a branching model.

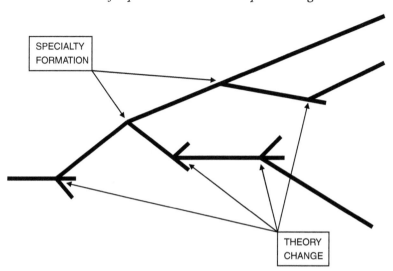

Figure 3 A diagram after Darwin: Kuhn's mature account of the development of science

regular contact with each other often have less in common to bring them together. They may no longer attend the same conferences. They may even begin publishing in different journals. They may also find commu-nication with each other more difficult (Kuhn 1969/1996, 177). But the process of specialization that Kuhn describes is essentially *epistemic* in nature. It is an epistemic shortcoming, the inability of a community of scientists to adequately model the domain of their field with the existing taxonomy that leads the community to divide and thus create a new spe-cialty. In fact, Kuhn claims that "specialization and the narrowing of the range of expertise [are] ... the necessary price of increasingly powerful cognitive tools" (1991a/2000, 98). Confined to the conceptual resources available in the existing scientific taxonomies, the resolution of some out-standing problems may not be possible. Certain phenomena may even evade our detection unless a radical change is made to an existing tax-onomy or lexicon. Thus, sometimes scientists must be prepared to relin-quish part of their current domain of study to those who are willing to employ a new modified lexicon better suited to the study of the recalci-trant phenomena.[4]

---

[4] Andrew Abbott (2001) also develops an account of the dynamics of scientific fields that attributes an important role to conceptual changes. Unlike Kuhn's account, Abbott's account is intended to apply to the social sciences only, and it is principally concerned with the dynamics *within* a field, rather than the dynamics that lead to the creation of new fields.

Importantly, Kuhn's account of specialization provides insight into the role that conceptual developments can play in the creation of a new specialty. But we should be mindful of an important insight of the second-generation sociological studies of specialization discussed earlier. The creation of a new specialty is a complex process, often involving cognitive, instrumental, and social changes. And separating the influence of the various factors can be challenging, especially when they are intertwined. For example, a social factor like crowding in a field, which may be caused by external factors of the sorts that Ben-David and Collins cite in the case of the development of experimental psychology, could play an important role in leading to the sorts of conceptual developments that would result in the creation of a new specialty. For as a scientific field gets crowded there may be more intense competition which can lead to or accelerate the process leading to a significant discovery.

As I mentioned above, philosophers are inclined to think of the creation of new specialties, and especially the resulting communication barriers they create, as a temporary state on our way to developing a unified science. Indeed, one committed to such a view might cite the fact that there are some specialties that are born at the intersection of two fields or sub-fields, for example physical chemistry or molecular biology. One might think, these cases give us reason to believe that science will be unified ultimately.

Kuhn, though, thinks otherwise. In fact, he explicitly discusses such cases, and he reaches a very different conclusion. He does recognize that both physical chemistry and molecular biology were "born at an area of apparent overlap between two existing specialties" (1991a/2000, 97). Further, he is aware that "at the time of its occurrence this [type] of split is often hailed as a reunification of the sciences" (97). But as far as Kuhn is concerned, "as time goes on … the new shoot seldom or never gets assimilated to either of its parents" (97). Rather, as the new field and the two fields from which it developed continue to develop "each of the fields [will have] a distinct lexicon" (98). Kuhn grants that "the differences [in the lexicons] are local, occurring only here and there" (98), but they can be significant enough to make communication between those working in the various specialties challenging. Thus, unification is not a reasonable expectation, according to Kuhn (98). "There is no lingua franca capable of expressing, in its entirety, the content of all or even any pair" of scientific fields (98). Indeed, Kuhn goes so far as to claim that the pursuit of unity may "well place the growth of knowledge at risk" (98).

These cases draw attention to an important part of Kuhn's account of specialization, one that justifies the comparison with biological speciation.

According to Kuhn, the motivation for creating a new scientific specialty is to improve the conceptual tools for understanding the world. And different conceptual tools are needed for modeling different phenomena.

In order to ensure that the appropriate conceptual tools are both developed and subsequently taken up by the relevant researchers it is imperative that the researchers are isolated from those in neighboring specialties. Such isolation provides the necessary barriers for the conceptual innovation to take hold, to become an integral part of the conceptual framework and research practices (see Kuhn 1991a/2000, 98; also Kuhn 1992/2000, 120). Thus, according to Kuhn, conceptual innovations are both the *cause* of barriers between specialties and require barriers between specialties if they are to develop. The "growing conceptual disparity between the tools deployed in ... two [neighboring] specialties" is "what keeps them apart and leaves the ground between them as apparently empty space" (120). Kuhn described the lexicons used in neighboring specialties as incommensurable, comparing the relationship between the lexicons in neighboring specialties to the relationship between competing theories during a revolution. But, as we saw in chapter 4, this is a distinct type of incommensurability, differing in important ways from what is now called "meaning-incommensurability." Most importantly, the incommensurability that arises between the lexicons in neighboring fields serves a constructive epistemic function, allowing the scientists working in the two fields to develop concepts, practices, and instruments suited to the phenomena specific to their field.

This notion of incommensurability also seems to undermine the goal of developing a unified science. If the lexicons in different specialties use the same terms equivocally, then there will be no single lexicon fit for all sciences. In invoking the notion of incommensurability in this context, Kuhn draws attention to the value of models and theories developed for a narrow range of phenomena. Such specialized theories, though narrow in scope, promise to deepen our understanding of the world. No single set of concepts and models can do justice to the range of phenomena scientists seek to understand. The fragmentation that results is the price we must pay for the depth of understanding afforded by specialization.

SUPPORT FOR KUHN'S ACCOUNT: VIROLOGY AND ENDOCRINOLOGY

One of the most valuable features of Kuhn's account of specialization is that it shows how conceptual developments in science can contribute to the creation of new specialties. By emphasizing the role of conceptual developments in the creation of new specialties, he draws attention to the epistemic dimension of the process. Kuhn, though, does not provide examples from the history of science to support his account. My aim in this section is to analyze two cases that support his account of specialization, specifically, the creation of the field of endocrinology and the creation of the field of virology.

The creation of the field of endocrinology provides a clear illustration of the process Kuhn describes. R. A. Gregory (1977) explains that:

[T]he discovery in 1902 by Bayliss and Starling ... of the duodenal hormone secretin was ... a signal event in the history of physiology. A simple experiment ... revealed that the functions of the body were normally co-ordinated not only by the nervous system, but also by the mediation of specific chemical agents formed in, and transmitted from, one organ to others by way of circulation, conveying a message intelligible only to those cells equipped to capture the "chemical messenger" and decipher the encoded instructions for modification of their activity. By the discovery of what came to be called "hormones" there was opened a new era of physiology, the beginning of endocrinology as we know it today. (105)

Before Bayliss and Starling made their discovery, physiologists generally assumed that the functions of the body were coordinated by the nervous system, and the lexicon physiologists worked with reflected this way of understanding the body. At that time, physiologists lacked the conceptual means to account for chemical messengers traveling through the bloodstream. Consequently, in making their discovery, Bayliss and Starling needed to invoke a new concept, "hormone" or "chemical messenger." Importantly, in this particular case, Bayliss and Starling were not motivated by the desire to accommodate a persistent anomaly. Rather, it was not until they discovered that some functions of the body were not coordinated by the nervous system that there was an anomaly.

Because Bayliss and Starling's conceptual innovation was at odds with the prevailing understanding of physiologists, their discovery significantly altered the field of physiology. "Hormone" became the key concept in the new field of study, endocrinology. But despite the emergence of this new specialty, the traditional study of physiology was, to a large degree, left very much intact as it was before the discovery of hormones. Physiologists continued to study the various functions of the body that were coordinated by the nervous system. One significant change that the discovery of hormones had on physiology, though, was to narrow the domain of the field. The revolutionary discovery of hormones reduced the range of phenomena that physiologists were expected to explain in terms of the operation of the nervous system. Physiologists relinquished responsibility for modeling and understanding functions of the body that were not coordinated by the nervous system. For example, they relinquished responsibility for explaining those newly discovered functions of the body that involved hormones.

Investigating the various functions of the body that are coordinated by "chemical messengers" requires different conceptual tools and laboratory practices from those then available to physiologists. The scientific study of these functions is sufficiently different from the scientific study of those that are coordinated by the nervous system to warrant the attention of specialists.

It is important to recognize that rather than modifying the existing taxonomy to accommodate the concept "hormone," physiologists *could* have tried to explain the observed phenomena in terms of the then operative lexicon. But had they tried to do this it is likely that an accurate understanding would have eluded them. At any rate, it is likely that some phenomena would have remained inexplicable, set aside as anomalies to be dealt with later. The existing lexicon, built on the assumption that the functions of the body are coordinated by the nervous system, was not fit to explain the phenomena under consideration.

This example nicely illustrates Kuhn's view of specialization. A significant discovery required radical changes to the taxonomy of a field with the result that a new field was born and the domain of the original field was subsequently truncated. The cognitive tools that were designed for one set of purposes, the study of the nervous system in the functioning of the body, were discovered to be inadequate for the range of phenomena to which they were *assumed* to apply. A new specialty was thus created to overcome this problem.

The creation of virology as a separate field of study also illustrates the process of specialty formation described by Kuhn. In the late nineteenth century and the first half of the twentieth century what we now call viruses were often studied as anomalous phenomena by various scientists working in a variety of specialties, including bacteriology, pathology, and biochemistry (Waterson and Wilkinson 1978). Some scientists attempted to understand viruses as microbes. Working with the conceptual and experimental resources of bacteriology, they were perplexed by the problem of growing viruses in artificial media. For these bacteriologists, this was an important anomaly (see van Helvoort 1994, 186). Other scientists attempted to understand viruses as non-organic entities. Working with the conceptual and experimental resources of chemistry, they were concerned with understanding how a toxic non-living substance could replicate (see Hughes 1977, 79–84, 90). For these biochemists, this was an important anomaly. But in the 1950s "a fundamental conceptual change occurred with the recognition that all viruses share a common structure despite observed differences in morphology, host specificity and pathological activity" (Hughes 1977, 100; see also van Helvoort 1994, 216). It was then realized that viruses were a distinct kind of phenomenon and could not be adequately conceptualized as either an organism or as a toxin. Thus, a new taxonomy or lexicon had to be constructed to accommodate "virus" as a kind.

The lexicon *bacteriologists* used before the discovery of viruses remained essentially the same after the discovery. Subsequent research in bacteriology was thus more or less continuous with the practice before the discovery of viruses. Bacteriologists, though, were able to relinquish responsibility for explaining certain phenomena that they had previously regarded as their responsibility, phenomena that had, until then, remained anomalous. A similar situation occurred in pathology and biochemistry, the other specialties in which scientists investigated the phenomena that came to be recognized as viruses.

These two episodes in the history of science illustrate the important role that conceptual changes can play and have played in the creation of new scientific specialties. Sometimes a new specialty is created as a result of a significant discovery. Such a discovery sometimes leads a community of scientists to split the domain of their field and form two separate research communities, each pursuing their research with a taxonomy or lexicon suited to their different needs and interests. The creation of a new specialty is thus one means by which scientists are able to enrich their understanding of the world. In fact, some phenomena are apt to elude our understanding until scientists make changes to an existing lexicon. Such changes allow scientists to approach the study of recalcitrant phenomena with new conceptual tools. And when a lexical change gives rise to a new specialty, scientists in the parent specialty realize the limits of their model, and thus relinquish responsibility for explaining phenomena they were ill-equipped to explain.

ANTICIPATING CRITICISM

In light of the sociological studies of specialization discussed earlier, I anticipate two criticisms of Kuhn's account of specialization. I will present the criticisms as they apply to the case of virology, though similar criticisms could be made with respect to the creation of endocrinology as well. My aim in this section is to address these criticisms in an effort to offer additional support for Kuhn's account of specialty formation, and in particular the role he attributes to conceptual innovations in the process.

First, one might argue that it was the introduction of new instruments and techniques that was responsible for the creation of virology rather than the taxonomic change outlined in the previous section. In support of this view one can cite the fact that the discovery of viruses depended upon a series of developments made in filtering, a process that played an integral role in enhancing scientists' understanding of viruses. One can also cite the fact that the electron microscope also played an indispensable role in the discovery of viruses (see Hughes 1977, 96 and 98; Waterson and Wilkinson 1978, 105–06).

Second, one might argue that virology was created, not as the result of conceptual changes, but rather as a result of crowding in existing fields. In support of this claim, one can cite the fact that even before 1900 Martinus Beijerinck had developed a concept of a non-cellular life form to account for the agent that causes the tobacco mosaic disease, an agent that would subsequently be identified as a virus (see Hughes 1977, 48–51). Since the required conceptual developments seem to have been available long before the creation of virology as a specialty in the 1950s, one could conclude that the creation of virology as a specialty depended, ultimately, on the migration of physicists into the biological sciences. Much like the young physiologists who created experimental psychology because of the bleak prospects for employment in physiology, these physicists were seeking new career opportunities and created them by applying the methods of physics to biological phenomena (see Hughes 1977, 91).

I will now address these criticisms, beginning with the first. It is undeniable that technological developments played an important role in the discovery of viruses, as do such developments in many discoveries. But I believe it is a mistake to regard the technological developments as the ultimate cause of the creation of virology as a specialty. There are two considerations that lead me to believe this. First, rather than being an epistemic asset, at times the reliance on techniques was an impediment to the discovery of viruses. Before scientists developed the modern concept of virus, technique-determined characteristics, like filterability and microscopic invisibility, were often used to identify substances as "viruses" (see van Helvoort 1994, 202). But, as Sally Hughes notes, such "technique-determined physical characteristics ... were inadequate criteria by which to differentiate viruses from all other types of infectious agents" (1977, 87; see also van Helvoort 1994, 202). That is, these characteristics do not systematically distinguish viruses from other superficially similar but unrelated phenomena. According to Hughes, what scientists lacked was "information about [the] intrinsic biological properties [of viruses]" (Hughes 1977, 87). Consequently, I believe it is a mistake to attribute the creation of virology either entirely or principally to developments in instrumentation.

In addition, I believe that it is a mistake to contrast technological or instrumental developments with epistemic developments. Developments in instrumentation are often tied to conceptual developments. This is because developments in instrumentation often either (1) depend upon conceptual developments or (2) occur simultaneously with conceptual developments. Kuhn illustrates this point in his discussion of the discovery of x-rays. In developing the technology to detect x-rays, Roentgen simultaneously discovered x-rays (see Kuhn 1962a/1996, 56–57). Making a sharp distinction between developments in instrumentation and conceptual developments, as some of the earlier sociological studies did, is thus misleading.

Let us now consider the second criticism, that the specialty of virology was created as a result of crowding in an existing field or specialty. I have three replies to this criticism. First, I believe that the criticism is based on a Whig reading of history. It is only with hindsight, with the knowledge that viruses form a distinct kind of entity, that we identify all the various investigations that led to the discovery of viruses as contributions to virology. Scientists involved in these investigations saw things quite differently. Before the 1950s, when scientists used the term "virus" it was applied indiscriminately to a variety of phenomena, many of which have subsequently been identified as bacteria (Hughes 1977, 92–95).

Augustine Brannigan (1981) makes a similar point with respect to Mendel's alleged contribution to genetics. Although *now* many regard Mendel as the father of genetics, Mendel did not see his research as either an attempt to create such a field or as a contribution to such a field. We mislead ourselves and undermine our efforts to understand science if we regard all research that ultimately contributes to the discovery of viruses as directed toward that end.

Second, though there were earlier anticipations of the modern concept, like Beijerinck's proposal which "agrees to some extent with the modern definition of the virus" (Hughes 1977, 57), it is a mistake to say that he discovered the same thing that was discovered in the 1950s. The vague understanding that Beijerinck had was, at the time, (1) in conflict with widely accepted background assumptions, (2) criticized by others, and (3) not well supported by the then available data (Hughes 1977, 57–58; Waterson and Wilkinson 1978, 27–30). Waterson and Wilkinson note that a number of Beijerinck's contemporaries who were researching the tobacco mosaic disease, including C. J. Koning, K. G. E. Heintzel, Friedrich Hunger, and Adolf Mayer, did not even understand "Beijerinck's flight of thought" (1978, 27). Others, including Emile Roux, Dmitri Ivanovski, and Eugenio Centanni, were critical of Beijerinck's conceptual innovation (27–28). Hence, this is not a case in which the conceptual work was more or less complete, and all that was needed to create the new specialty was to enlist a sufficient number of dedicated, hard-working scientists.

Third, many of the scientists who were ultimately responsible for the creation of virology as a specialty were not looking for new career opportunities, as were the nineteenth-century physiologists who helped create the field of experimental psychology. Many of them were well-established researchers, with secure positions, and actively involved in research in a number of *specialties*. Hence, the pioneers in virology lacked the incentives to create a new social niche that had motivated the early founders of experimental psychology. Their attraction to the field was more due to the fact that they saw opportunities to make discoveries in a hot topic area.

In summary, contrary to what is implied by the objections outlined at the beginning of this section, a comprehensive explanation for why virology came to be created as a new specialty in the 1950s requires reference to both (1) conceptual developments in our understanding of viruses, and (2) the accompanying taxonomic or lexical changes required to articulate an adequate understanding of the phenomena. Consequently, Kuhn's account of specialization offers significant insight into the dynamics of specialty formation.

SPECIALIZATION AND COMMUNICATION

Specialization has important implications for the spreading of scientific knowledge. With the creation of a new scientific specialty the expertise of individual scientists narrows. Consequently, scientists are increasingly required to depend upon and trust the findings of those working in other specialties. This increasing dependence that characterizes contemporary science is one of the principal costs of specialization in science. Indeed, it is concerns such as this that have led philosophers to believe that the increasing specialization that characterizes modern science is unfortunate.

As new specialties are created, communication barriers are created and these can sometimes be impediments to the development of science. Some important discoveries have depended upon scientists transcending the barriers created by specialization. For example, Paul Thagard (1999) notes that the discovery of the bacterial theory of ulcers required knowledge from two different medical specialties, pathology and gastroenterology (84–85, 181). Apparently, much of the knowledge needed to make the discovery was available for some time before the discovery was made. But it was only when two scientists from different specialties worked collaboratively that the discovery was made. Thagard claims that it is unlikely that either scientist would have made the discovery working alone. Recognizing this predicament, there have been attempts to develop computer software that can identify connections between bodies of scientific literature from different fields that might lead to scientific discoveries (see Swanson and Smalheiser 1997).

But it is important to recognize that the communication barriers created by specialization do not have only negative effects on science. Full communication within the community of scientists would be detrimental, for effective lines of communication not only serve to spread the truth; they also serve to spread falsehoods. Kevin Zollman (2007) claims that "in some contexts, a community of scientists is, as a whole, more reliable when its members are less aware of their colleagues' experimental results" (574). Using computer simulations and the resources of game theory, Zollman considers three model communities of scientists: (1) a community in which "agents are ... only connected with those on either side of them," (2) a community much like (1) except that "one of the agents is connected to everyone else," and (3) a community "where everyone is connected to everyone" (578). His computer simulations show that greater connectivity in a community "corresponds to faster but less reliable convergence" (580). So, as paradoxical as it might sound, "in many cases a community made up of less informed individuals is more reliable at learning correct answers" (575). Hence, insofar as specialization prevents full communication in the larger community of scientists, it serves the important function of blocking the spread of error.

The threat of the communication and spread of falsehoods in science should not be underestimated. In a study of the fate of retracted articles in scientific journals, Budd *et al.* (1998) found that many retracted articles continue to be cited frequently after they have been retracted. Budd *et al.* identified 235 retracted articles published between 1966 and 1997.

These articles "received a total of 2034 postretraction citations" (297). A citation counted as a postretraction citation if the citing article was published at least one year after the retraction was published (see Budd *et al.* 1998, 296). This way of operationalizing the term allows scientists enough time to see the retraction before they cite the article. Examining a sub-set of these postretraction citations, they found that most citations did not acknowledge that the article had been retracted.

In his modeling of communication in science, Zollman (2007) found that in our efforts to advance the goals of science we may have to trade off accuracy for speed (583). The same features of a research community that make for efficient communication, social density for example, can be impediments to reliability (581–82, 584). And as Zollman notes, not surprisingly, there is no obvious means to determine how such trade-offs are to be made. Such decisions will, undoubtedly, be influenced by the non-epistemic values of those making the decisions, for in some cases we may prefer efficient communication over reliability, whereas in other cases we may privilege reliability.

The sorts of barriers that make communication between scientists challenging not only serve to limit the spread of error. These barriers ensure that research communities can develop concepts that are suited to modeling the phenomena they study without too much interference from scientists in neighboring specialties. Although scientists often rely on other scientists in neighboring fields for their expertise on matters relevant to their own research, as a research community, in their efforts to realize their research goals, scientists working in a specialty often need to develop the concepts they use in ways that are incommensurable with the way similar concepts are used in neighboring specialties.

In 1982, Barry Barnes, a key proponent of the Strong Programme, could rightly claim that Kuhn had very little to say about scientific specialization (see Barnes 1982, 14). But, as Kuhn developed his view, he developed the rudiments of an account of specialty formation. In his later writings, he came to argue that the formation of a new scientific specialty is an alternative response to crisis and anomalies in a scientific field.

My aim in this chapter has been to present and defend Kuhn's account of specialization. I have argued that Kuhn provides an answer to a key question left unanswered by earlier sociological studies of specialization: how can conceptual or epistemic developments in science contribute to the development of a new scientific specialty? As we saw above, sometimes scientific discoveries will lead a research community to split into two research communities, each working with a lexicon suited to its needs and interests. This is what we saw in both the creation of the field of endocrinology in the early 1900s and the creation of the field of virology in the 1950s.

Specialization has been shown to be an alternative response a community of scientists makes when confronted with persistent anomalies. That is, rather than seeking to accommodate the anomalies by replacing the accepted theory with a new, alternative theory, a specialty sometimes splits, with some scientists pursuing the study of the anomalous phenomena with new conceptual tools designed for that purpose, and others working with the older theory in a restricted domain. Kuhn's account of specialty formation, with its focus on conceptual developments as the driving force, also makes clear why this dimension of scientific change is relevant to philosophers of science.

As noted in the previous chapter, some critics and commentators have taken issue with Kuhn's biological metaphors that underlie his evolutionary epistemology. I have sought to show that these are in fact fruitful metaphors, providing valuable insight into understanding the dynamics of scientific change. With that said, it should be acknowledged that Kuhn does make some mistakes in his discussions of biological evolution. For example, he claims that the comparison of scientific evolution and biological evolution breaks down because there is a destructive element in scientific evolution that is not present in the process of biological speciation (Kuhn 1992/2000, 120). Kuhn has in mind here the fact that older concepts are discarded as new scientific theories are developed (120). But, contrary to what he suggests, biological evolution is equally destructive. Many biologists have remarked on how wasteful the biological world is. In some species, for example, very few offspring survive into adulthood. Waste and destruction are thus as prevalent in the biological world as they appear to be in the scientific world. More work, though, needs to be done in order to determine how far the biological metaphors can be applied.

CHAPTER 8

Taking stock of the evolutionary dimensions of Kuhn's epistemology

It is worth taking stock of the various respects in which Kuhn's epistemology of science is aptly called an *evolutionary* epistemology, for, as we saw earlier, the term "evolutionary epistemology" does not identify a single, well-defined approach to epistemology.

In Part I, I argue that Kuhn believes there are genuine revolutionary changes in science. But, throughout Part II, I have argued that Kuhn's epistemology is an evolutionary epistemology. One might think there is a tension in Kuhn's epistemology of science with his characterization of science in both evolutionary and revolutionary terms. The question is: can Kuhn have it both ways? Can an epistemology of science be both an evolutionary epistemology and also acknowledge that there are revolutionary changes in science? In answering this question I aim to clarify what in Kuhn's epistemology is aptly characterized as revolutionary and what is aptly characterized as evolutionary.

Indeed, some suggest that scientific change can be described as *either* evolutionary in nature *or* revolutionary in nature, *but not both*. Stephen Toulmin (1970/1972), for example, defends an evolutionary theory of scientific change, contrasting it with Kuhn's revolutionary account of scientific change. Toulmin suggests that insofar as Kuhn's account of scientific change can be construed as evolutionary at all, it is a form of "catastrophism" (1970/1972, 47). Contrary to what Kuhn suggests, Toulmin claims that even the most significant changes that occur in science can be better accounted for in terms of "mechanisms of variation and perpetuation" (45), and thus do not involve the sorts of radical discontinuities that we associate with the term "revolution." Hence, Toulmin believes philosophers of science must choose between an evolutionary account of scientific change and a revolutionary account.

The impression that an epistemology of science is either evolutionary or revolutionary is also reinforced by the way Popper's view is sometimes contrasted with Kuhn's view. Sometimes when Popper's view is compared with Kuhn's, Popper is represented as offering an evolutionary account of scientific change and Kuhn is represented as offering a revolutionary account of scientific change. Those who contrast Popper's view and Kuhn's view in this way also seem to assume that an epistemology is either evolutionary or revolutionary. But this characterization of their views is problematic for a number of reasons.

Both Popper and Kuhn explicitly acknowledge that some aspects of scientific change are aptly characterized as revolutionary and some aspects are aptly characterized as evolutionary. At the London conference in the mid 1960s, Popper and his followers portray Popper as encouraging perpetual revolution. And Popper objects strongly to Kuhn's characterization of normal science, arguing that "the 'normal' scientist, as Kuhn describes him, is a person one ought to feel sorry for" (1970, 52). Popper claims that, from the perspective of his own critical philosophy of science, "the 'normal' scientist … has been taught badly … He has been taught in a dogmatic spirit: he is a victim of indoctrination" (52–53). Popper thus recognizes that there are revolutionary changes in science. And, similarly, Kuhn was unequivocal in characterizing his own view of scientific change as evolutionary in nature. Responding to his critics in London, Kuhn claims that his "view of scientific development is fundamentally evolutionary" (1970b/2000, 160).

I believe that Kuhn's characterization of his epistemology of science as both evolutionary and revolutionary is consistent. I will show this by explaining the respects in which Kuhn believes scientific change is revolutionary, and then by explaining the respects in which he believes scientific change is evolutionary.

In Kuhn's developed view, it is changes of theory, or episodes of theory replacement, that are characterized as revolutionary changes. Such changes are revolutionary because they involve the replacement of an accepted scientific lexicon with a new lexicon that divides the phenomena into different categories. And these changes to a new lexicon are not mediated with the help of robust, shared standards that provide scientists with the means to make unequivocal judgments about which of the competing lexicons, the old or the new, is superior. Thus, just as in political revolutions, scientists engaged in episodes of theory change have no agreed-upon standards to enable them to resolve their differences.

The answer to the question "in what respects is Kuhn's account of scientific change evolutionary?" is more complex. We have seen in the previous three chapters a number of respects in which it is appropriate to describe Kuhn's epistemology as an evolutionary epistemology.

First, Kuhn believes that we must adopt an evolutionary perspective when we study science. The evolutionary or historical perspective requires us to see science as an activity or process underway. Scientists accept a body of beliefs and they evaluate new theories and hypotheses against that background. There is no Archimedean platform or foundation outside our accepted theories on which to stand in an effort to evaluate competing hypotheses.

Second, Kuhn believes we need to develop a non-teleological account of scientific change. An adequate understanding of science will continue to elude us as long as we conceive of science as directed toward uncovering the truth. Indeed, contrary to the traditional philosophical perspective on science, Kuhn believes that there is very little that we can explain if we regard science as directed toward developing a true account of reality. This has been a crucial part of Kuhn's epistemology since the publication of *Structure*. And he was quite attuned to the fact that such a radical change in perspective would meet with fierce resistance. Indeed, he felt that Darwin's theory met with its fiercest resistance on exactly this point. Nonetheless, Kuhn remained deeply committed to the necessity of making this change in orientation in our efforts to develop an epistemology of science.

Third, Kuhn increasingly came to realize that not all crises in science are resolved in the same manner. Not all crises end with the replacement of a once-long-accepted scientific lexicon by a new scientific lexicon. Sometimes a crisis is resolved by a process that resembles speciation. The research community divides, with one part continuing to work with the older, long-accepted lexicon, and the other part developing a new lexicon suited to the study of hitherto unstudied or misunderstood phenomena. Importantly, Kuhn insists that specialization is not merely a temporary condition that scientists must accept on their way to developing a unified science. Rather, the process of specialty creation, he claims, is the key means by which scientists have developed increasingly accurate models of the phenomena they seek to understand.

Thus, contrary to what Toulmin implies, the revolutionary and evolutionary aspects of Kuhn's epistemology of science fit together. In synthesizing the revolutionary and evolutionary aspects of Kuhn's epistemology, it is useful to distinguish between (1) Kuhn's account of scientific change and (2) Kuhn's recommendations on how to study science. Consider the issue of scientific change first. According to Kuhn, scientists resolve crises in their fields in one of two ways: one way is aptly characterized as a revolutionary change and the other way is aptly characterized as an evolutionary change, involving a process similar in many respects to speciation in the biological world. Consider, now, the issue of the perspective from which we should examine science. The perspective the epistemologist of science must adopt is an evolutionary perspective, seeing scientific change as a process underway and as a process that is pushed from behind, rather than directed toward some true account of nature, waiting to be discovered by us. This perspective is in no way at odds with revolutionary changes in science.

Given the evolutionary dimensions of Kuhn's epistemology outlined above, his view is aptly regarded as a version of anti-realism. Kuhn believes that it is a mistake to explain the success of science in terms of the truth of our theories. He is quite explicit that we have little reason to believe that the various theories developed in different specialties present a unified or consistent account of the structure of the world. Rather, the value of our theories is a function of the fact that they serve the local needs of the scientists working in the fields in which they were developed and are subsequently employed.

Anti-realism also fits with the revolutionary dimensions of Kuhn's account of theory change. If successive theories in a scientific field really do posit fundamentally incompatible ontologies, then certain forms of scientific realism seem untenable. The more recently developed theories in a field are, no doubt, superior to those that preceded them. They afford us all sorts of benefits, including increased accuracy. But these gains are not a consequence of us getting ever closer to the truth. In fact, given that the development in scientific fields involves a sequence of theories being replaced by incompatible theories, there is little basis for claiming that there is a convergence on anything.

Indeed, John Worrall (1989) claims that the existence of scientific revolutions in the history of science is the greatest challenge to most forms of realism (103). Worrall, though, is critical of anti-realism. He opts for what many regard as the weakest form of realism, structural realism. According to the structural realist, though there may be little or no continuity in the ontologies of successive theories in a field, scientists do sometimes retain the mathematical equations used in earlier theories. The retention of these equations suggests that scientists have got at the underlying relations between the phenomena they study, even if their beliefs about the entities posited by their theories are mistaken (119). Thus, any philosopher of science prepared to admit that there are revolutionary changes in science of the sort Kuhn claims exist seems to be faced with a choice between (1) accepting some form of anti-realism and (2) accepting a very weak form of realism, like structural realism.

PART III

Kuhn's social epistemology

Sociologists of science are certainly correct to see the importance of the social dimensions of scientific inquiry in Kuhn's work and the constructive role he attributes to such factors in aiding scientists in realizing their epistemic goals. In this respect, sociologists have been more sympathetic and careful readers of Kuhn's work than philosophers. But, to many philosophers, sociologists of science have taken things too far. Their emphasis on the social dimensions of science leave one wondering what role rationality and evidence play in resolving disputes in science.

In Part III, I want to examine the respects in which Kuhn's epistemology of science is a social epistemology. I also want to show that Kuhn's social epistemology does not threaten the rationality of science. Consequently, his social epistemology provides a useful framework for developing an epistemology of science.

I begin by examining the charge that Kuhn is a constructionist. This is an especially troubling accusation, in part because constructionism is such a poorly defined view. Indeed, as we will see, constructionism has come to mean many things, and some of them pose no threat to the epistemic integrity of science. Hence, my aim is to clarify the nature of Kuhn's constructionism. I will show that Kuhn's form of constructionism is not threatening to the rationality of science.

I continue to clarify the relationship between Kuhn's views and those of the sociologists most influenced by him, the advocates of the Strong Programme. There are important respects in which their views differ dramatically. Clarifying the differences between Kuhn's view and that of the advocates of the Strong Programme will be vital to our developing an appreciation of (1) Kuhn's epistemology of science, as well as (2) the differences between the sociology of science and the philosophy of science. I argue that Kuhn's form of nominalism differs in a significant respect from the Strong Programme's nominalism. Whereas the proponents of the Strong Programme believe that *every* act of classification is underdetermined, Kuhn makes a weaker claim, specifically that there is no unique way to group the things in the world. Thus, Kuhn thinks the mind-independent world places more constraints on how scientists carve nature up into kinds than the proponents of the Strong Programme acknowledge. I also argue that Kuhn is an internalist. Contrary to what many philosophers think, Kuhn believes that disputes in science are resolved on the basis of evidence.

In chapter 10, I examine the senses in which Kuhn's epistemology of science is aptly described as a *social* epistemology. I identify four issues that Kuhn addresses that make his epistemology a social epistemology. First, Kuhn examines the socialization process that scientists-in-training undergo which enables them to see the world as they need to in order to make a contribution to science. Second, Kuhn believes that knowledge is produced by groups of scientists rather than individual scientists. The type of social group that figures most importantly in Kuhn's social epistemology is the scientific specialty or research community. Third, Kuhn believes that theory change is a form of social change. More precisely, theory change, which involves the replacement of a long-accepted scientific lexicon with a new, incompatible lexicon, also involves significant social changes in the research community. And, as far as Kuhn is concerned, we will not understand the nature of theory change until we develop a better understanding of the social dimensions of the process. Fourth, Kuhn believes that philosophers of science need to draw on work in the sociology of science if they are to develop an adequate understanding of the social dimensions of theory change. Indeed, social scientific research into the nature of social groups and social change will be valuable resources as we seek to better understand the epistemic culture of science.

In chapter 11, I aim to demonstrate the value of empirical investigations of the process of theory change. My point of departure is Kuhn's claims about young scientists as the source of new theories and old scientists as a source of resistance to new theories. I show that these claims of Kuhn's are not supported by the available data. I then examine a case of theory acceptance in a research community, with special attention to the social changes that occurred in the research community. I focus on an examination of the social processes underlying the acceptance of the theory of plate tectonics in geology in the 1960s. The data on which I report suggest that the new evidence gathered in support of the theory of plate tectonics played a crucial role in the acceptance of the new theory. I also argue that more empirical studies are needed if we are to develop an adequate descriptive account of how science works.

In Part III I aim to correct a number of misconceptions about Kuhn's views on scientific inquiry, scientific change, and scientific knowledge. In addition, I provide a compelling case for why an epistemology of science must be a social epistemology, one concerned with understanding the social dimensions of science.

CHAPTER 9

Kuhn's constructionism

Many of Kuhn's critics think that Kuhn is the philosopher most responsible for emphasizing the impact of non-cognitive factors in science, and thus threatening the rationality of theory choice (see Lakatos 1970/1972; Laudan 1984; Boghossian 2006). Both the sociologists of science who were inspired by his work and many of his philosophical critics regard Kuhn's view as a form of constructionism.[1] But as Ian Hacking (1999) notes, "constructionism" connotes different things to different people. Many sociologists and historians of science self-consciously approach their subject as constructionists (see Latour and Woolgar 1986; Shapin 1996, 9–10; Golinski 1988/2005). Many philosophers of science, on the other hand, believe that constructionism entails a pernicious form of relativism (see, for example, Brown 1989, 37; Boghossian 2006, 118–22). Hence, for many philosophers of science calling someone a constructionist amounts to a refutation of their view.

In this chapter, I aim to clarify the relationship between Kuhn's view of science and constructionism. In clarifying the nature of Kuhn's constructionism, I also aim to clarify Kuhn's relationship to both philosophy of science and sociology of science.[2]

In *The Social Construction of What?*, Hacking (1999) offers a detailed and subtle analysis of constructionism with respect to science,

---

[1] Hacking rightly notes that "most items said to be socially constructed could be constructed only socially, if constructed at all. Hence the epithet 'social' is usually unnecessary" (1999, 39). Hereafter, I will drop references to the "social," acknowledging that, unless otherwise stated, constructionists are *social* constructionists.

[2] Kuhn thought of *Structure* as a book for philosophers. It was initially published by invitation in the logical positivists' *Encyclopedia of Unified Science*. But his relationship with philosophers was often strained. He was initially denied his promotion to *full* professor by the philosophy department at Berkeley (see Kuhn 2000b, 301–02). His relationship with sociologists was also strained. He felt misunderstood by them, though many sociologists of science claimed to be building on his view of science (Kuhn 1977b, xxi; 1992/2000, 110). Kuhn's relationship with philosophers was further strained by the fact that some sociologists, in particular the proponents of the Strong Programme, claimed to be inspired by Kuhn.

distinguishing between three distinct constructionist theses. Hacking claims that Kuhn strongly endorses all three theses associated with constructionism. Consequently, Hacking's analysis provides a useful starting point for understanding Kuhn's relationship to constructionism. Further, Hacking has long been one of the most sympathetic commentators on Kuhn.

I argue that although Hacking is correct to regard Kuhn's view as a form of constructionism, he misrepresents Kuhn's view in two respects. First, contrary to what Hacking claims, Kuhn endorses only a weak form of nominalism. According to the extreme nominalism endorsed by the Strong Programme in the Sociology of Scientific Knowledge, *every* single act of classification is underdetermined. This is not Kuhn's view, though he does believe that there is no ultimate best way to classify things, no single way prescribed by nature. Second, Hacking is mistaken in claiming that Kuhn believes that *external* factors are the cause of consensus in science. Rather, I aim to show that Kuhn believes that *epistemic* factors are responsible for stabilizing belief in science. Many philosophers have been misled to think that Kuhn is an externalist, I argue, because they misunderstand the role Kuhn attributes to non-epistemic factors in theory change. He does believe that non-epistemic factors play an important role in helping a research community resolve the problem of theory choice. But non-epistemic factors do not *determine* which theory is chosen. Hence, Kuhn is not an externalist. By clearing Kuhn of the charge of externalism, I aim to show that he does provide a viable framework for developing an *epistemology* of science. In keeping with traditional philosophy of science, Kuhn's concerns are with reasons and evidence, though he recognizes the important role that the social structure of research communities play in enabling scientists to realize their epistemic goals. Thus, once we get a clear understanding of the nature of Kuhn's constructionism, it should be clear that his view poses no great threat to the epistemic integrity of science. His view is based on a deep appreciation of the value of science.

Kuhn is also frequently accused of supporting some form of relativism. He adamantly rejected that characterization of his view. I examine three versions of the charge of relativism that have been raised against Kuhn, including (1) the claim that Kuhn believes that subjective factors determine theory choice, (2) the claim that Kuhn's conception of rationality is subjective, and (3) the claim that Kuhn gives us no reason to believe that science is a particularly effective way to investigate the world. All of these charges, I argue, are ungrounded.

KUHN THE CONSTRUCTIONIST

I want to begin by examining Hacking's characterization of Kuhn's constructionism. Hacking argues that Kuhn endorses three theses widely
accepted by constructionist sociologists of science: the contingency thesis, nominalism, and externalism. Before proceeding, readers should be
cautioned about two matters. First, in a chapter devoted to social construction and science, Hacking presents these three constructionist theses, making clear that he regards Latour and Woolgar (1986) and Andrew
Pickering (1984) as typical constructionist sociologists of science. It is only
at the end of the chapter that Hacking briefly discusses Kuhn's constructionism. And in a lighthearted manner Hacking notes that on a scale
of one to five, where five represents an extreme constructionist and one
represents a non-constructionist, Kuhn ranks five with respect to all three
theses. Hence, my evidence that Hacking regards Kuhn as a constructionist of the most extreme sort is based on Hacking's ranking of Kuhn's view
with respect to the three theses. Second, I am not principally concerned
with determining whether Hacking is correct in the way he characterizes
Kuhn's constructionism, but rather with clarifying the nature of Kuhn's
constructionism. Hacking's analysis is useful because it helps us see the
complexity of the debate about constructionism. And it is important to
clarify the nature of Kuhn's relationship to constructionism because some
of his critics seem to feel that they have shown that Kuhn's account of science can be dismissed merely by showing that he is a constructionist. This
is a line of criticism that I wish to contest.

Let us now examine Hacking's characterization of Kuhn's constructionism. First, Hacking claims that constructionists endorse a contingency thesis according to which "alternative 'successful' science is in
general always possible" (Hacking 1999, 69). Hacking claims that those
who endorse the contingency thesis believe, for example, that modern
physics might have developed in a direction that did not involve the concept of a quark. It is undeniable that physics could have developed in
different directions. The various historical contingencies that shaped the
direction of research in physics could have been different, leading to the
development of different sorts of physical theories from those we currently accept. The constructionist, though, maintains that some of the
possible ways in which physics might have developed would have resulted
in a physics that is as successful as our contemporary physics. That is,
these alternative physical theories would have led to fruitful discoveries
and afforded our effective interaction in the natural world. Indeed, they

would have afforded us successes equal to those our current theories have afforded us. According to Hacking, this is Kuhn's view.

Hacking contrasts the contingency thesis with a view he calls "*inevitablism*," the thesis that the various sciences, insofar as they are successful, must develop in more or less the way they have in fact developed. Inevitabilists believe that the alternative paths of change that science might have taken would have resulted in a science less successful than contemporary science.

Second, Hacking believes that constructionists are nominalists, and thus deny "that nature has joints to be carved" (Hacking 1999, 83). Hacking seems to be of two minds about what nominalism entails. In his contribution to *World Changes*, Hacking claims that "nominalism is not to be confused with 'name-ism,' the doctrine that things have nothing in common except their names" (1993, 307n. 5). But in *The Social Construction of What?* Hacking claims that "nominalism is a fancy way of saying name-ism. The most extreme name-ist holds that there is nothing peculiar to the items picked out by a common name" (1999, 82). Given that my chief concern is to understand Kuhn's constructionism, we need not determine which of Hacking's characterizations of nominalism is correct. I will focus on Hacking (1999).

According to constructionists, scientists can carve nature in different ways, grouping the same set of things in numerous different ways. Constructionists emphasize that no particular way of grouping things has a special claim on the allegiance of scientists. That is, no way of grouping things is aptly described as *the true* or ultimate way to group things. This, Hacking claims, is Kuhn's view. Hacking contrasts nominalism with a view he calls "inherent structuralism," the view that the way things ought to be ordered and grouped by scientists is ultimately *determined* by nature. On this view, there is a proper way to group things, and changes in theory are our attempts to get closer to that ultimate and proper way of grouping things. Ultimately, what we seek are theories that embody concepts that cut nature at its joints.

Third, Hacking claims that constructionists are externalists. "Externalism" and "internalism" are used to pick out a variety of different contrasting views. In the sociology and history of science an internalist believes, roughly, that changes in science are to be explained in terms of features internal to science, like developments in theory or conceptual innovations. Steven Shapin suggests that internalists are inclined to view "science as 'essentially theory'" (1992, 342). In contrast, externalists believe that some changes in science are best explained in terms of the influence of factors external to science, like innovations in commercial or military practices, for example. Herbert Butterfield's (1957/1965) *The Origins of Modern Science* is regarded as a typical internalist history of science, tracing the development of *ideas* in early modern science, whereas Shapin and Schaffer's (1985) *Leviathan and the Air-Pump* is regarded as a typical externalist study of science. On their account, it was not inevitable that the experimental philosophy associated with Robert Boyle and the Royal Society came to dominate in seventeenth century England. Rather, the influence of various external factors secured the victory of the experimental program in early modern England.

Shapin (1992) rightly notes that the distinction between internal and external is harder to draw than many acknowledge. One might wonder, for example, whether innovations in the organization of a research team are to count as an internal or an external factor. On the one hand, such innovations have an impact on scientists' abilities to realize their research goals, and thus seem aptly characterized as internal. But on the other hand, because the organization of a research team involves social factors, it may seem apt to characterize it as external.[3] We need not draw the line between the internal and the external too precisely, for it is far from clear that there is a principled way to distinguish internal and external factors. The important point for our purposes is that insofar as philosophers are concerned with the issue, they are concerned to defend a brand of internalism according to which when scientists are faced with competing theories, the accepted theory is accepted because of the influence of the *epistemic merits of the theory*. For philosophers, the alternative seems to be some form of skepticism.

Externalists thus believe that the available data underdetermine theory choice. Consequently, when there is a dispute between advocates of competing theories the sorts of factors that ultimately bring about consensus in the research community are "social factors, interests, [and] networks" (Hacking 1999, 92). That is, factors like interests, rather than evidence, ultimately lead scientists to reach a consensus about which of two competing theories is superior. External factors thus bridge the gap between

[3] Helen Longino (2002) has objected to the common contrast between the social and the rational, arguing that something can be both social and rational. Shapin (1992) raises a similar concern (349).

theory and data. According to Hacking, this is Kuhn's view on consensus formation in science.

The competing view, the view that philosophers of science generally endorse, is that *reasons* and *evidence* stabilize belief in disputes in science. Because scientists respond to evidence, adjusting or modifying their beliefs to new data, consensus will emerge as the scientists working in a field attend to new data, with each scientist individually recognizing the merits of the theory deemed to be superior.

It is because Hacking believes that Kuhn endorses the contingency thesis, nominalism, and externalism that he regards Kuhn as a constructionist. Hacking refers to these three claims as the "sticking points" that divide constructionists and their critics. Importantly, Hacking believes that these are genuine "philosophical barriers [between constructionists and their critics], real issues on which clear and honorable thinkers may eternally disagree" (1999, 68). Hence, Hacking doubts that constructionists and their critics will ever resolve their differences.

It is worth noting that Hacking's characterization of constructionism is deeply philosophical, as he relates constructionism to persistent and long-standing philosophical debates. Sociologists and historians are apt to focus on different issues in their understanding of constructionism. For example, constructionist sociologists of science would be more inclined to examine the role social factors, like gender and power, play in science (see Sismondo 1996, chapter 4). And historians regard constructionism as "a methodological orientation [rather] than a set of philosophical principles" (Golinski 1988/2005, 6). It involves the recognition that "scientific knowledge is a human creation ... rather than simply the revelation of a natural order that is pre-given and independent of human action" (6).

KUHN'S NOMINALISM

In the next two sections I want to examine the extent to which it is appropriate to call Kuhn a constructionist. In this section, I argue that Hacking is correct in claiming that Kuhn (1) endorses the contingency thesis and (2) is a nominalist. But I argue that Kuhn endorses a weaker form of nominalism than the form of nominalism that Hacking attributes to him. Contrary to what Hacking claims, Kuhn *does* believe that nature has joints. What makes Kuhn a nominalist, I argue, is his conviction that there is no single proper way to group things with our theories.

Hacking is certainly correct to characterize Kuhn as a nominalist of sorts, though Kuhn (1993/2000, 229) objects to the variety of nominalism that Hacking attributes to him in Hacking's contribution to *World Changes*. Kuhn took Hacking to be attributing to him the view that "there are real individuals out there, and we divide them into kinds at will" (Kuhn 1993/2000, 229). It is as if there are *no* constraints on how we group things. This is a misrepresentation of Kuhn's view.[4]

Hacking, though, seems to characterize the nominalist dimension of Kuhn's view correctly when he claims that Kuhn believes that:

[T]he world is a world of individuals; the individuals do not change with a change of paradigm [that is, a change of theory] ... But ... the world in which we work is a world of kinds of things ... all working is under a description ... Descriptions require classification, the grouping of individuals into kinds. And that is what changes with a change of paradigm [that is, with a change of theory]. (Hacking 1993, 277)

Like constructionism, "nominalism" has come to denote a variety of different positions. It was originally a *metaphysical* view about the nature of universals. But as Alexander Bird (2003) notes, there is now a position that is aptly described as *epistemological* nominalism. Whereas the metaphysical nominalist "rejects the *existence* of universals," the epistemological nominalist "rejects *knowledge* of universals" (Bird 2003, 705; emphasis added). We will see that insofar as Kuhn is a nominalist he is an epistemological nominalist, for, like the positivists, his concerns are epistemic, not metaphysical (see Sharrock and Read 2002, 66).

Of the three constructionist theses outlined above, nominalism is the one that figures most importantly and explicitly in Kuhn's view of science. Kuhn's commitment to nominalism is most evident from his remarks on the nature of theory change in his later writings (see, for example, Kuhn 1979b/2000, 205; 1991a/2000, 104). We saw above that when Kuhn recognized the ambiguity in the term "paradigm" he came to describe theory change as involving a taxonomic or lexical change, a change in the way scientists group the things in the world. For example, as we saw earlier, Ptolemaic astronomers did not regard the Earth as a planet, whereas Copernicans insisted that the Earth was like Jupiter and Mars, merely a planet. Indeed, with this change in groupings came a change in meaning (see Kuhn 1962a/1996, 128; 1979b/2000, 205; 1991b/2000 218). "Planet"

---

[4] In an effort to respond to Hacking's characterization of his view, Kuhn (1993/2000) claims that some of the referents of some terms scientists use, "like 'force' and 'wave front'," are not aptly construed as individuals (229). I think Kuhn's concern here is tangential to the real issue that separates his own view from the view that Hacking attributes to him.

came to mean a celestial body that orbits the sun, whereas previously the term connoted "wandering star," that is, a star that does not move with the fixed stars. Similarly, after Darwin, ornithologists reordered the relations between bird species, grouping species together that were not grouped together before (see Andersen *et al.* 2006, 87–88). This change in groupings was also accompanied by a change in meaning. Since Darwin, "species" is a term that relates to lineages. As far as Kuhn is concerned, every change in theory involves both changes in the way things are grouped and changes in the meaning of key terms. Indeed, Kuhn chastises Hacking for aiming to "eliminate all residues of a theory of meaning from [Kuhn's] position" (Kuhn 1993/2000, 229).

Kuhn also maintains that there is no point in talking about an ultimate way of grouping things, a way that cuts nature at *the* joints (see Hoyningen-Huene 1989/1993, 267). He explicitly denies "that nature has one and only one set of joints to which the evolving terminology of science comes closer and closer with time" (Kuhn 1979b/2000, 205). Indeed, as we saw in chapter 6, even in *Structure*, he encourages us to give up the image of science as an "enterprise that draws constantly nearer to some goal set by nature in advance" (1962a/1996, 171). The history of science does not support such a view. The sequence of changes in theory in a field does not reveal a pattern of convergence (in this regard, see Laudan 1984). But this is not to deny that "the number of solved scientific problems and the precision of individual problem solutions ... increase ... markedly with the passage of time" (Kuhn 1977c, 320). Indeed, Kuhn never doubted that science was successful in these respects (see Kuhn 1979b/2000, 206).

Kuhn's brand of nominalism, though, must be distinguished from other more extreme forms of nominalism, for example, the view of Barnes *et al.* (1996), which they call "finitism."[5] In an effort to understand Kuhn's nominalism, it is worth clarifying how his view differs from the view of Barnes *et al.*, according to whom "there is an ineliminable indefiniteness associated with [applying a concept]" (54). In fact, on their view, *every* new application of a concept is underdetermined (54).[6]

---

[5] Interestingly, Hacking claims that it is a mistake to characterize the advocates of the Strong Programme who explicitly claim to be influenced by Kuhn as constructionists (Hacking 1999, 65). I disagree with Hacking about this. Bird's discussion of finitism and its relationship to Kuhn's view is especially insightful (see Bird 2000, 218–25), though I think Bird is mistaken to regard *their* view as a metaphysical thesis. They are sociologists of scientific *knowledge*; their concerns are *epistemic*.

[6] Some of the critics of the Strong Programme claim that the advocates of the Strong Programme believe that "anything can count as obeying the rules" with respect to applying concepts (see, for example, Brown 1989, 37). This is a mistaken interpretation of their view. Clearly, the prevailing

Barnes *et al.* distinguish their own account of classification from a conventionalist account of classification. According to the conventionalist view, once a community settles on conventions of classification, the conventions

determine our subsequent taxonomic activity. It is as if the world is a cake, ready to be cut in any number of ways, indifferent to how it is actually sliced. We decide on how to do the slicing, but once it has been done the status of everything in the world is fixed, and we must subsequently proceed as the conventions "require." (Barnes *et al.* 1996, 55)

According to this account of classification, what is underdetermined is the choice between conventions or ways of classifying things. But, once this matter is settled, once a means of classification is selected, classification becomes a straightforward procedure. Scientists just need to decide which natural differences they will attend to in their theories. This characterization of the conventionalist account is similar to the view that Rudolf Carnap (1950) defends, a view that Michael Friedman (2001) and others suggest has strong affinities to Kuhn's view.

Barnes *et al.* believe that this popular account of classification is mistaken. They argue that "reflection on the empirical process of classification suggests a better metaphor [than the conventionalist metaphor outlined above]. We have put our knife into the cake and cut a certain way. But *nothing determines* how we should continue to cut" (Barnes *et al.* 1996, 55; emphasis added). Hence, according to the finitist, settling on a set of conventions does not eliminate ambiguity, for every instance of classification is underdetermined. This seems to be the view that Hacking attributes to Kuhn (see Hacking 1999, 99; see also Bird 2003, 706). Indeed, Barry Barnes (2003) also suggests that Kuhn endorses their finitist account of "concepts, norms, rules and laws, [and] model and theories" (129).

But Kuhn's account of classification differs from both of these accounts. On the one hand, Kuhn does not believe that classification is as underdetermined as finitists claim. In fact, as noted above, Kuhn says explicitly that Hacking is mistaken when he attributes to him the view that "there are real individuals out there, and we divide them into kinds *as we will*" (1993/2000, 229; emphasis added). Kuhn does believe that the choice between competing theories or lexicons *can* be underdetermined. But

conventions can underdetermine how a concept is applied without thereby making every application of a concept correct. Brown's interpretation of their view implies that they believe "anything goes," a slogan made popular by Feyerabend, though even his use of that expression is often misunderstood.

once a theory is widely accepted, once a research community is engaged in normal science, the choice of how to classify objects is *generally* not underdetermined. It is in this respect that Kuhn's account of classification differs from the finitists' account.

On the other hand, Kuhn does not believe that classification is ever as settled as the conventionalist suggests. After all, Kuhn believes that even in normal science the classification of a particular object or event can be underdetermined. Kuhn calls such cases "anomalies." And as far as he is concerned, anomalies are an inevitable part of *any* normal-science tradition. Indeed, anomalies are what ultimately undermine a normal scientific research tradition. Hence, unlike the conventionalist, Kuhn believes that no scientific taxonomy will ever unambiguously determine the classification of *all* objects and events. Part of the reason Kuhn is led to believe this is that he believes that scientists inevitably encounter objects that cannot be adequately accounted for given the conceptual resources of the accepted theory.

One might think that compared with the finitists Kuhn hardly deserves to be called a nominalist. After all, finitists suggest that the world does not have *any* joints at all; instead, the world is like an undifferentiated cake. It can be carved in an indefinite number of ways. Although Kuhn acknowledges that the world can be carved in many different ways, he insists that once scientists settle on a set of principles for carving, that is, on a set of concepts, there are real constraints on how they classify objects in the future, constraints imposed by *mind-independent* features of the world (see Kuhn 1991a/2000, 101; 2000b 317; see also Kuhn 1977c, 331). Indeed, Kuhn claims that "the metaphors of invention, construction and mind-dependence are ... misleading ... The world is not invented or constructed" (1991a/2000, 101). Rather, as Kuhn explains, we "find the world already in place ... It is entirely solid: not in the least respectful of an observer's wishes and desires; quite capable of providing decisive evidence against invented hypotheses which fail to match its behavior" (101).

Thus, it is a mistake to think that Kuhn believes that nature has *no joints*. Kuhn, though, does believe that there is no unique set of joints that our concepts must get at in order to be useful or afford successful interactions in the world (see Kuhn 1979b/2000, 205). It is for this reason that it is appropriate to describe him as a nominalist. Compared with the finitists, though, we should probably call him a *weak* nominalist, something less than a "five" on Hacking's scales of constructionism.

In his early work, Kuhn gives two reasons for believing that no theory, that is, no set of principles for carving the world, will last forever. First, as far as he is concerned, every theory is partial, leading scientists to attend to *some* features of the world, but not others (see Kuhn 1962a/1996, 126). Indeed, it is because the theory they accept and work with restricts their vision that scientists working in a normal-science tradition are so effective at problem-solving. Such a partial perspective focuses scientists' attention narrowly and selectively (Kuhn 1962a/1996, 64). Kuhn suggests that scientists may not even notice certain phenomena when they are working within a particular theoretical framework. And novices, not yet committed to *any* theory, typically do not know what phenomena they ought to attend to.

And because all theories are partial, a change of theory often involves a change in what scientists seek to understand or model. For example, late Renaissance astronomers who accepted the existence of crystalline spheres did not ask what keeps the planets in their orbits. But by the late 1570s, after the sighting of a comet that was determined to have passed through the space beyond the Moon, the existence of such spheres came to be questioned, and it became a legitimate question. Similarly, new questions and phenomena became important when geologists accepted the theory of plate tectonics. For example, once tectonic plates were hypothesized to exist, geologists became interested in studying how these plates could move laterally.

Second, scientists' interests change as they develop the accepted theory in their efforts to answer hitherto unsolved problems. Hence, they inevitably discover that a set of concepts that proved useful in understanding one set of phenomena is less useful in understanding other phenomena. A new theory is thus needed in order to account for the phenomena that eventually come to be of interest to scientists.

Given Kuhn's nominalism, it is not surprising that he also endorses the contingency thesis, the view that science need not have developed the way it did in order for it to be as successful as it is. If there is no ultimate proper way to group things in the world, then it seems that science could develop in a number of different ways. Further, it seems that more than one of the possible ways science could have developed may have led scientists to develop a theory that enables them to realize their goals and manipulate the world in predictable ways.

Incidentally, Kuhn does believe that the general developmental pattern of mature sciences is *inevitable*. In mature fields, periods of normal science are interrupted by crises that end in revolutions which begin new phases of normal science. But in conceding that this developmental pattern is inevitable, Kuhn does not mean to suggest that the *conceptual* developments in a successful field necessarily follow a specific sequence. Hence, Kuhn would deny that the sequence of physical theories from Aristotle's to Descartes', from Descartes' to Newton's, and from Newton's to Einstein's was inevitable.

Inevitabilists, though, do draw attention to an important insight. Some discoveries can only be made after other specific discoveries have been made. Hacking gives the example of Lagrange and his contemporaries, who had to develop the "mathematics in order to discover certain consequences of Newton's laws of motion and gravitation" (1999, 76). Without these developments in mathematics, these consequences of Newton's theory may never have been known. Hence, as inevitabilists suggest, certain discoveries can only be made on the condition that other specific discoveries have already been made. But one can grant this and still deny that the path we actually took in science is the only possible path to a successful science.

KUHN'S INTERNALISM

We saw above that Hacking is correct in claiming that Kuhn endorses the contingency thesis and is a nominalist of sorts. Hacking, though, is mistaken in calling Kuhn an externalist. Indeed, there seems to be a lot of confusion surrounding the relationship between Kuhn's view and externalism and internalism. The sort of externalism that concerns us here is the claim that social factors and interests are what bring about consensus in science. More precisely, social factors and interests *determine* the outcome of disputes between advocates of competing theories. This is not Kuhn's view (see Kuhn 2000b, 287).

Interestingly, Hacking recognizes that Kuhn does not regard himself as an externalist. Hacking claims that Kuhn "insisted that he ... [is] an internalist historian of science, concerned with the interplay of ideas, not the interactions of people" (1999, 43; see also Hacking 1993, 282). Still, Hacking maintains that Kuhn is an externalist (1999, 99). And other philosophers share Hacking's view on this issue. Ironically, Kuhn notes that *historians* "sometimes complained that [his] account of scientific development is too exclusively based on factors *internal* to the sciences themselves" (Kuhn 1977b, xv; emphasis added). Sociologists of science hold a similar view. They claim that Kuhn became increasingly internalist in orientation after the publication of *Structure* (see Pinch 1979, 439). But, apparently, Alexandre Koyré claimed that Kuhn had struck the right

balance, and "brought the internal and external histories of science …
together" (Kuhn 2000b, 286).

As we look at the evidence for treating Kuhn as an internalist it is
important to remember that not all social influences on science threaten
the integrity of science. For example, the external factors that influence
what topics are researched do not necessarily threaten the *epistemic* integrity of science. Many have noted that more money and research efforts
have been spent trying to understand the causes and cures of diseases that
tend to affect white middle-class men than those that tend to affect work-
ing-class women. This is largely a function of the power and resources
each group commands in society. But we can admit that such factors
influence the *direction* of scientific research without thereby suggesting
that choices between competing theories are determined by factors external to science. It is the latter sort of influence that concerns philosophers.
Hence, if Kuhn is an externalist, then he must maintain that *theory choice*
is determined by external factors.

Kuhn (1968/1977) acknowledges that non-scientific social factors often
influence the *rate* of change in science (119). Further, he claims that external factors have a more pronounced influence in underdeveloped fields.
But he insists that in mature fields, the focus of his concern, research communities are largely insulated from the influence of external factors (119).
Consequently, in such fields, internal factors are responsible for change.

Before explaining in detail why Kuhn is not an externalist, I want to
briefly consider why many have been led to think otherwise, for Hacking
is not alone in regarding Kuhn as an externalist. Many have been led to
think of Kuhn as an externalist because of his remarks on how *subjective* factors influence scientists when they are faced with a choice between
two competing theories (see Kuhn 1977c). Kuhn *does* in fact believe that
individual scientists are influenced by subjective factors in their decision-
making. For example, he claims that subjective factors may lead one
scientist to weigh the five commonly identified objective criteria of theory choice – accuracy, simplicity, consistency, scope, and fruitfulness –
differently from another scientist. As a result, confronted with a choice
between two competing theories, one scientist might prefer the simpler
theory to the theory that is broader in scope, when another scientist is led
to choose the theory that is broader in scope (Kuhn 1977c, 322; see also
Kuhn 1979b/2000, 204). Such divergences in judgments could even happen when scientists have access to the same body of data. But this admis-
sion on Kuhn's part, this recognition of the underdetermination of theory

choice by evidence, does not warrant calling him an externalist. After all, Kuhn is not saying that subjective factors are responsible for *consensus formation* in science.

According to Kuhn, when theory choice is underdetermined by the evidence, subjective factors ensure that different theories are *developed* (see Kuhn 1977c, 332). As a result, the various competing theories are refined. In the process, new evidence is amassed, and, in time, the epistemically superior theory emerges as the victor (see Hoyningen-Huene 1992, 493 and 496). Importantly, this process enables the research community to determine which theory is superior. When scientists are influenced by subjective factors, it induces different scientists to work with different theories, which ensures that competing hypotheses are developed to the point where it becomes clear which theory is superior. Subjective factors thus provide a research community with a means to resolve the problem of theory choice. Such factors cause individual scientists to take sides in a dispute which divides research efforts effectively, thus aiding the research community in selecting the superior theory. Kuhn notes that "what from one viewpoint may seem the looseness and imperfection of choice criteria … appear as indispensable means of spreading the risk which the introduction or support of novelty always entails" (1977c, 332). Hence, on Kuhn's view, subjective factors merely play an *instrumental* role in ensuring that the superior theory is the one that is ultimately accepted by the *research community*.

Kuhn uses an episode from the history of science to illustrate his point about the role of subjective factors. In 1600, an astronomer could justifiably choose to endorse either Ptolemy's theory, Copernicus' theory, or Brahe's theory. None of the three theories was more accurate than the others (see Kuhn 1977c, 323). But by the 1640s the situation had changed dramatically. The new evidence generated and collected and the refinements made to the Copernican theory rendered it vastly superior to the competitor theories. These developments were made possible because individual astronomers were influenced by subjective factors earlier, when it was less apparent which theory was superior. Even Kepler's neo-Platonism may have played a constructive role in this way. Kuhn claims that "if Kepler or someone else had not found … reasons to choose heliocentric astronomy [the] improvements in accuracy [he achieved] would never have been made, and Copernicus's work might have been forgotten" (Kuhn 1977c, 323). But the subjective factors that influenced scientists are not what determined which theory was ultimately accepted by the research community. That is, Kepler's neo-Platonism is not what persuaded other astronomers to accept the Copernican hypothesis. And it is certainly not what led to the consensus when the Copernican theory became the dominant theory.[7]

In summary, according to Kuhn, subjective factors play an important role in the process of theory change, influencing individual scientists in choosing a theory when theory choice is underdetermined. But this does not make Kuhn an externalist. After all, as mentioned in chapter 1, a single scientist choosing one theory over another does not constitute an instance of theory change, at least not in the sense relevant to philosophers of science concerned with the epistemology of science. Theories are sustained and transmitted by research communities (Kuhn 1991b/2000, 220–21 [cf. 1977b, xx]; see also Hacking 1993, 276).[8] And a theory is not aptly described as the accepted theory unless it is widely held in the community.

Having clarified what Kuhn has said about the role subjective factors play in science, I will now explain why Kuhn is not aptly described as an externalist. Kuhn believes that *epistemic factors* stabilize belief in the research community. It is in this respect that he is aptly described as an internalist. Kuhn notes that "before the group accepts [a new theory, it] has been tested over time by the research of a number of men, some working within it, others within its traditional rival" (1977c, 332). A consensus will only emerge and the competitor theories will only be abandoned when one theory is deemed to be epistemically superior to the competitors. Granted, different scientists are apt to be moved by different considerations in their assessment of the competing theories (see Kuhn 1977c, 329). But, at some point, most agree that the conceptual resources afforded by one theory are superior to those afforded by the competitors.

---

[7] One of Kuhn's main lasting contributions to philosophy of science is his recognition that the influence of subjective factors cannot be relegated to the context of discovery. Such factors can and often do play a role in the context of justification (see Kuhn 1977c, 326–27, for his concerns about relegating the influence of subjective factors to the context of discovery). This insight has been developed by numerous philosophers of science since Kuhn. To some extent, David Hull's (1988) invisible hand explanation for the success of science, Philip Kitcher's (1993) work on the division of epistemic labor in research communities, and Miriam Solomon's (2001) "social empiricism" build on this dimension of Kuhn's project.

[8] Although Kuhn believes that the research community is *the locus of theory change*, he is not suggesting that research communities are *agents*, capable of choosing between competing theories. In fact, Kuhn explicitly denies this (see Kuhn 1991a/2000, 103; 2000b, 283; see also Sharrock and Read 2002, 47). Some commentators and critics, including Hoyningen-Huene, have failed to draw this distinction (see Hoyningen-Huene 1992, 495; 1989/1993, 200 and 205).

Moreover, Kuhn claims that belief in science is also *de*stabilized by epistemic factors, not social factors and interests. It is a misfit between the accepted theories and the world that leads scientists to develop alternative theories and introduce conceptual innovations. Kuhn explains that "alterations in the way scientific terms attach to nature ... come about in response to pressures generated by observation or experiment" (1979b/2000, 204). Copernicus, for example, was motivated to develop his theory of planetary motion because he was dissatisfied with Ptolemy's appeal to the equant, a device he thought was employed in an ad hoc fashion. Copernicus sought to achieve a better fit between theory and world. Similarly, Kepler's appeal to elliptical orbits was an attempt to achieve a better theory-to-world fit in an effort to account for Brahe's data on Mars. In both cases, the impetus for theory change is *internal* to science. Hence, contrary to what Hacking claims, Kuhn is an internalist.

RATIONALITY AND RELATIVISM

In explaining why Kuhn is an internalist, I have also provided some insight into why critics are mistaken in accusing Kuhn of irrationalism. Were it the case that a research community settles disputes on the basis of subjective factors, then Kuhn would be guilty of making theory choice a non-rational process. But since Kuhn makes it clear that such matters are settled on the basis of epistemic considerations, this version of the charge of irrationalism is ungrounded.

I want now to examine two additional charges of relativism that have been leveled against Kuhn, specifically, the charge that his conception of rationality is subjective and the charge that he gives us no reason to believe that science is a more effective way to gain knowledge of the world than other practices. Both of these charges, we will see, are ungrounded.

The fact that Kuhn had such an impact on the development of the Strong Programme has affected the way he is commonly read by philosophers. Because the proponents of the Strong Programme are relativists, some have regarded Kuhn's view as entailing some form of relativism. And because the Strong Programme's relativism is thought to entail irrationalism, Kuhn's view is also sometimes characterized as undermining the rationality of science. In the remainder of this chapter, I want to examine and address one recent version of this line of reasoning, a line of reasoning developed by Michael Friedman.

Friedman (2001) has recently taken issue with Kuhn's view of rational-ity, arguing that it leads to relativism (51). He notes that the proponents of the Strong Programme who claim to be building on Kuhn's work openly embrace relativism (49). Friedman recognizes that Kuhn rejects the "rela-tivist implications of his views" and has sought to distance himself from the views of the Strong Programme. But Friedman claims that Kuhn is unsuccessful in addressing the charge of relativism (50). Friedman traces the failure of Kuhn's view to what he regards as an inadequate conception of rationality. Kuhn believes the operative notion of rationality in science is instrumental rationality. Friedman, though, argues that "instrumental rationality is in an important sense private and subjective" (55). Because "human ends … are notoriously diverse and variable … there can be no ground for a truly universal rationality within purely instrumental rea-son" (55).

Friedman contrasts instrumental rationality with communica-tive rationality, a notion he borrows from Jürgen Habermas. Whereas the former type of rationality is merely concerned with "maximiz-ing our chances of success in pursuing an already set end or goal," the latter type of rationality is more substantive (see Friedman 2001, 54). Communicative rationality "is essentially public or intersubjective. It aims, by its very nature, at an agreement or consensus based on mutually acceptable principles of argument or reasoning shared by all parties in a dispute" (55).

Friedman's description of communicative rationality is elusive. Clearly, it is meant to be more substantive than instrumental rationality, but more substantive in what ways it is difficult to determine. In *After Virtue*, Alasdair MacIntyre gives us some sense of what a more substantive notion of rationality might involve. He claims that reason, conceived in a non-instrumental sense, "instructs us both [on] what our true end is and [on] how to reach it" (MacIntyre 1981/1984, 53). Instrumental rationality, on the other hand, is merely concerned with the latter task. It is not exactly clear how a substantive conception of scientific rationality would instruct us on what our ends are. But, minimally, Friedman's more substantive notion of rationality is meant to be less subjective than instrumental rationality.

Friedman grants that Kuhn is correct in claiming that "the scientific enterprise as a whole has in fact become an ever more efficient instrument for puzzle-solving" (2001, 53). That is, he grants that it is rational by the standards of instrumental rationality. But Friedman claims that even if science satisfies the standards of instrumental rationality, the threat of relativism persists. For, according to Friedman, one needs to show "that the scientific enterprise thereby counts as a privileged model or exemplar of rational knowledge of … nature" (53). Thus, in order to address this threat of relativism, one must show that science is the privileged way of investigating nature.

With respect to these two charges of relativism, it seems that Friedman considers Kuhn guilty merely because of his influence on the Strong Programme. Because the proponents of the Strong Programme are both self-proclaimed relativists and self-identified Kuhnians, Kuhn is branded as a relativist. In the remainder of this section, I want to defend Kuhn's view by addressing Friedman's concern about grounding the epistemic superiority of science, as well as his concern about the subjective and private nature of instrumental rationality. In addition to showing that these charges of relativism against Kuhn are ungrounded, I want to show that such charges are even unfair to the proponents of the Strong Programme. They are relativists of sorts, but they do not believe that other knowledge-seeking practices are as epistemically effective as science.

Let us first consider the Strong Programme. Contrary to what many of their critics suggest, the proponents of the Strong Programme recognize the success of science and place great value on the methods of science. Steven Shapin (1996), for example, claims that "science remains … certainly the most reliable body of natural knowledge we have got" (165). And Barnes *et al.* (1996) employ the methods of science in their sociological studies of science, thus demonstrating their commitment to the value of science. It would be very odd to employ the methods of science, as these sociologists of science do, if one did in fact have such a low opinion of their value, as their critics suggest they do.

There are two sources of confusion that have led many philosophers to assume that these sociologists of science doubt the efficacy of science. First, as both Barnes *et al.* (1996) and Shapin (1996) note, some think that to examine science objectively, that is, to adopt a "non-evaluative approach" to the study of science, is to be critical of science (Barnes *et al.* 1996, x; Shapin 1996, 165). But this is not so. In fact, as they note, scientists aim to adopt such an approach in their own studies. Hence, these sociologists of science merely approach their study of science with the same attitude scientists adopt in their studies. That is, they are merely being scientific.

Second, it seems that some philosophers think that the Strong Programme's symmetry principle implies that science is no better than other cultural practices. This is not so. The symmetry principle demands that we seek similar causes for both true beliefs and false beliefs (Barnes and Bloor 1982). In particular, it discourages us from explaining those beliefs that we now *think* are true as caused by reality, and explaining those beliefs that we deem to be false as caused by the intervention of some sort of social factor. The critics of the Strong Programme thus assume that social factors have a distorting effect on science. Consequently, the critics believe that if we seek to find the social causes of all our beliefs, then we are assuming that they are all contaminated. The proponents of the Strong Programme, though, do not assume that the impact of social factors on science is necessarily bad or distorting. A belief can be true even if it has a social cause.

Further, contrary to what their critics suggest, the symmetry principle does not imply that non-scientific practices are as efficacious as our best scientific practices. Rather, what the symmetry principle implies is that science should be investigated in the same way that other social practices are investigated. This methodological assumption, though, makes no presumptive judgment that science is either equal or inferior to other epistemic practices. And, given that the proponents of the Strong Programme embark on scientific studies of science, it is doubtful they believe that there are any better approaches or methods they could have chosen. The methods they employ in their own studies thus demonstrate their allegiance or commitment to science.

Kuhn also does not question either the efficacy of science or the epistemic superiority of scientific inquiry. Indeed, numerous commentators regard Kuhn as an apologist for science, taking for granted the success of science (see, for example, Fuller 2000; Barnes 2003, 135). And, like the proponents of the Strong Programme, Kuhn believes that empirical studies of science are the key to developing a better understanding of science.

Let us now consider Friedman's claim that instrumental rationality is private and subjective. Friedman believes that an instrumentalist conception of rationality like Kuhn's makes the goals of science private and subjective. Because advocates of such a conception of rationality regard the goals of science as private and subjective, they have no basis for claiming that science is getting at a true account of the world (see Friedman 2001, 54).

Again, Friedman is mistaken in the way he characterizes Kuhn's view. Given Kuhn's views about scientific research, it would be misleading to describe the goals and values of science as subjective or private. An individual scientist certainly does not have the liberty to determine the goals and values of her research field. Rather, scientists work in and as part of a research community or scientific specialty. Consequently, a scientist's research goals and values are determined, to a large extent, by the research community to which she belongs. In this respect, the goals of science are not aptly described as subjective or private. Granted, when there is a crisis in a field the members of a research community will be divided about what their goals should be. But it is difficult to see how this sort of situation can pose a deep or persistent threat to the rationality of science. And it is even more difficult to see how it might lead one to think that science is on an equal epistemic footing with other non-scientific practices.[9]

My aim in this chapter has been to clarify the nature of Kuhn's constructionism, for much of the resistance to his view is based on the fact that he is alleged to be a constructionist. Hacking has identified three distinct issues that divide constructionists and their critics – contingency, nominalism, and externalism – thus drawing attention to the complexity involved in classifying anyone as a constructionist. Hacking regards Kuhn as a constructionist in virtue of his position on all three of these issues.

Although I agree with Hacking that Kuhn is a constructionist, I have challenged Hacking's characterization of Kuhn's constructionism in two ways. First, I have argued that Kuhn is only a *weak* nominalist. He does not accept the strong nominalist thesis that nature has no joints. But Kuhn does believe that there is not a unique correct way to sort nature into kinds. Hence, he does not believe that a successful scientific theory need employ any specific set of kind terms. Second, I have argued that Kuhn is not an externalist. Contrary to what externalists claim, Kuhn believes that disputes between advocates of competing theories are resolved on the basis of a consideration of the *epistemic* merits and the promise of the theories, not on the basis of social factors and interests. Subjective factors do play an important role in the resolution of disputes, ensuring that competing theories are developed, and thus ensuring that the strengths and weaknesses of the theories are exposed. But, ultimately, consensus is achieved on the basis of epistemic considerations. Finally, I suggested

---

[9] Gerry Doppelt (1978) regards Kuhn's view as a form of relativism, but Doppelt claims that Kuhn regards reasons and their strength as *relative* to "the standards internal to particular paradigms [that is, theories]" (53). What counts as a reason is determined by the theory one accepts. This form of relativism, though, seems to pose no serious threat to the rationality of science, provided one is not trapped in a theoretical framework incapable of seeing an alternative view.

that, given that Kuhn is not an externalist, the common charge that he regards theory choice as either a non-rational or an irrational process is ungrounded. Hence, contrary to what some of his critics suggest, the mere fact that Kuhn is a constructionist does not constitute grounds for rejecting his view.[10]

---

[10] There is another way in which Kuhn's view might be and has been construed as a form of constructionism. Paul Hoyningen-Huene notes that Kuhn believes that the "phenomenal world" is "a world constituted by the activities of knowing subjects" (Hoyningen-Huene 1989/1993, 29). In this respect Kuhn's view is *similar* to Kant's view. Although Kuhn believes that the world is constituted by knowing subjects, Hoyningen-Huene adamantly insists that Kuhn is not an idealist. Hoyningen-Huene explains that "for Kuhn, reality, that is, a particular phenomenal world, is ... object-sided, independent of all influence by subjects, in its *substantiality*" (Hoyningen-Huene 1989/1993, 268). The similarities between Kuhn's view and Kant's view have been explored extensively by Hoyningen-Huene (1989/1993) and continue to be explored by Friedman (2001). We need not pursue this constructionist thread in Kuhn's work further here, though, for as James Conant and John Haugeland (2000) note, Kuhn came to repudiate certain Kantian elements that he had endorsed earlier in his career.

CHAPTER 10

What makes Kuhn's epistemology a social *epistemology?*

One of Kuhn's key contributions to the philosophy of science was to direct our attention to the epistemic relevance of the social dimensions of scientific inquiry. Kuhn shows us that there are limits to what we can learn about science and scientific knowledge when we restrict ourselves to a study of the logic of science, as the logical positivists and Popper do. Scientific inquiry is a complex social activity. And the social dimensions of science play an important role in ensuring the success of science. Kuhn's epistemology of science is thus a *social* epistemology of science. Kuhn, however, does not describe his project as a social epistemology of science.[1] This is not surprising, given that the term "social epistemology" became widely used among philosophers only in the 1980s, with the publication of the journal *Social Epistemology*.

The term "social epistemology" has come to mean different things to different people. Sometimes it connotes the study of such things as expertise or testimony as sources of knowledge (see Schmitt 1994, 4–17; Goldman 1999, chapter 4). At other times, social epistemology concerns science policy issues, including whether and to what extent the public which pays for science through taxation should shape the scientific research agenda (see, for example, Fuller 1999). And "social epistemology" sometimes connotes a concern with whether the social characteristics of inquirers affect their prospects of developing an objective account of the world or some part of it (see Schmaus 2008).[2] Kuhn's epistemology of

---

[1] Concern for understanding the social dimensions of science was a central part of Kuhn's project. When he was reflecting on the criticisms raised at the London conference in the 1960s, he remarked that "a new version of *Structure* would open with a discussion of community structure" (1970b/2000, 168). And in the Postscript to *Structure*, Kuhn explicitly notes that because "both normal science and revolutions are ... community-based activities ... to ... analyze them, one must unravel the changing community structure ... over time" (Kuhn 1969/1996, 179–80).

[2] R. K. Merton (1972/1973) uses the term "social epistemology" in an article in which he investigates the competing claims of epistemic superiority of outsiders and insiders (123n. 41). Many feminist epistemologists are also concerned with understanding how social marginalization can be an epistemic asset. But Merton's work has had no impact on the feminists who discuss this issue.

science is a social epistemology because he sought to understand how the social dimensions of science contribute to the success of science.

It was Kuhn's concern for the social dimensions of science that attracted the attention of sociologists of science, leading to the emergence of a new approach to the sociology of science, the Strong Programme's Sociology of Scientific Knowledge. The proponents of the Strong Programme were not content with studying the culture and institutions of science, the traditional subjects of sociological studies of science. Instead, they sought to understand how social factors influence the *content* of science.[3]

As I have repeatedly mentioned, Kuhn went to great lengths to distance his own view from the views of these sociologists who credit him with their inspiration. But many of Kuhn's philosophical critics treated and continue to treat his view as essentially indistinguishable from the new sociology of science, convinced that his project differs very little from it (Laudan 1984; Friedman 2001). And because the Strong Programme is perceived to threaten the authority of science, Kuhn's philosophy of science has often been deemed to undermine the epistemic authority of science. In this chapter, I want to clarify further the relationship between Kuhn's view and the views of the sociologists inspired by him. I will do this by examining the nature of Kuhn's social epistemology.

I focus on four key aspects of Kuhn's social epistemology. First, Kuhn analyzes the education of young scientists. He notes how uncritical the process is, and how it leads scientists to be myopic, attending to only certain features in the environment. But Kuhn claims that scientists are as effective as they are because of the socialization process they undergo. Second, Kuhn believes that scientific knowledge is produced by groups, indeed very special sorts of social groups. Consequently, an adequate epistemology of science must take this into account. Third, Kuhn believes

This is unfortunate, as Merton's analysis of the relative value of the two perspectives is quite insightful.

[3] Among sociologists of science, there is a common narrative of the development of their field. According to this narrative, the early 1970s were liberating times for sociologists of science. Prior to this time sociologists of science, in particular, the Mertonians, had been concerned mainly with studying the institutions of science as social institutions, leaving the study of the content of science to philosophers of science. But a careful reading of a number of the papers in Merton's (1973) *Sociology of Science* challenges this narrative. Merton was certainly concerned with issues that deserve to be classified as epistemological. His analyses of publication norms in science, for example, have profound epistemic implications (see Merton 1959, 1961/1973, 1963/1973). Merton suggests that science would never have become the important social institution it has become without scientists developing the means – scientific periodicals – and the inclinations – fed by peer recognition – that led them to make public the results of their research. This process began in the seventeenth century, when scientists sought to distinguish themselves from alchemists who were understandably extremely secretive.

that scientific change is a form of social change. Consequently, in order to understand science we must understand the nature of the social changes underlying changes of theory and scientific specialty formation. Fourth, Kuhn believes that philosophers must draw on social scientific research in an effort to develop an adequate epistemology of science. To neglect such research would impede us in developing an adequate understanding of science.

I end this chapter by briefly examining some of the normative implications of Kuhn's social epistemology of science.

MAKING SCIENTISTS

Kuhn's analysis of the education process in science is intended to uncover for the curious outsider, that is, the philosopher of science, how scientists are able to achieve their epistemic goals. In particular, he wants to explain how the scientists working in a sub-field are able to see the world the same way. This uniformity of vision plays a crucial role in making normal scientific research so effective.

Kuhn describes the education of young scientists-to-be as a highly regimented and controlled process (Kuhn 1962a/1996, 11; 1959/1977, 237). According to Kuhn, scientific education is "a relatively dogmatic initiation into a pre-established problem-solving tradition that the student is neither invited nor equipped to evaluate" (Kuhn 1963, 351). The process of enculturation into a scientific research community is remarkably uncritical. It aims to immerse the young trainees into a scientific culture, to enable them to see the world through the lenses of the accepted theories and to work with the accepted exemplars. The purpose of this process is to socialize the young scientists-in-training, to make them fit for scientific research. Indeed, according to Kuhn, science is so effective at realizing its goals because of this thorough socialization. This is what ensures that young scientists-in-training are able to see the sorts of things that their mentors and advisors are able to see so effortlessly, the sorts of things that they must learn to see in order to be effective researchers (see Kuhn 1962a/1996, 17).[4]

---

[4] This is the point that Norwood Russell Hanson (1958/1965) was making in his analysis of the perception of gestalt images. The well-trained scientist does not interpret the data. Rather, she sees the data as she was trained to see it, through the lenses of the accepted theory. Indeed, it is because perception is theory-laden that scientists are able to see as much as they can when they look at the world. When we compare what the scientist sees with what the layperson sees, it is as if the layperson has impaired vision. Lorraine Daston's (2008) work on scientific observation, in meteorology, for example, illustrates this same point.

Textbooks are explicitly designed to support the enculturation of young scientists-in-training. They are largely devoid of the history of the science, except to support a Whig account of the field, one that makes the present practices and accepted theories look like the inevitable culmination of all previous successes. The errors of the discipline's heroes are not mentioned, nor are their interests insofar as they diverge from the research interests of contemporary researchers in the field. It is only Newton the physicist, not Newton the alchemist or theologian that contemporary students of science learn about in their science textbooks.

The problems or exercises students tackle as part of their education make the student familiar with the accepted exemplars and scientific lexicon, the tools that are indispensable to their subsequently making a contribution themselves. They are designed to teach the student how key theoretical concepts relate to each other and the world.

But the socialization process that makes new scientists not only makes people capable of doing scientific research, it also makes people myopic and narrow in their expertise. Scientists are often unable to see things that will later prove to be important after a new theory is accepted. And scientists become specialists in a very narrow research area. Outside their area of expertise, scientists are dependent on the expertise of others, much like the typical layperson. Scientists who need to draw on research in neighboring fields have little choice but to accept what they read in authoritative sources or what they are told by other scientists working in the relevant fields.

SCIENTIFIC KNOWLEDGE IS A SOCIAL PRODUCT

The primary way in which Kuhn's epistemology of science has influenced developments in social epistemology is in drawing attention to the fact that scientific knowledge is produced by *groups*. This fact is not incidental to the success of science. Kuhn is explicit about this, claiming that:

[T]hough science is practiced by individuals, scientific knowledge is intrinsically a *group* product and ... the manner in which it develops will [not] be understood without reference to the special nature of the groups that produce it. (1977b, xx)

The role that groups play in scientific research is vividly illustrated in the following anecdote from James Watson's account of the discovery of the structure of DNA. Watson notes that at one point in the discovery process he uncritically consulted a respected reference book, J. N. Davidson's *The Biochemistry of Nucleic Acids*, in order to find information about tautomeric forms (Watson 1968, 182). Working with this information, Watson was led down a dead-end trail. Later, he discovered that the information in the text was widely regarded as false by crystallographers, thanks to the insight of Jerry Donahue (190). Watson was clearly outside his area of expertise. But, with Donahue's assistance, he was set in a new and more promising direction. Working alone, Watson was unlikely to discover the structure of DNA. But with the help of others, including Donahue, Francis Crick, Maurice Wilkins, and Rosalind Franklin, the structure of DNA was discovered. Watson's experience illustrates Kuhn's point that the products of scientific research are really the products of the specialized *groups* created to produce scientific knowledge. Kuhn believed that an epistemology of science must take account of these facts.

Throughout *Structure* Kuhn went to great lengths to show that the social structure of research communities plays an indispensable role in creating the type of social group equipped to realize the goals of science. He describes scientific research communities, especially in their normal scientific research phases, as tradition-bound and rather insular. The effectiveness of normal scientific research is due to the fact that the members of the research community can take for granted many assumptions about both how the world is structured and how science is properly done.

It is only when scientists settle their differences about the fundamentals in a field that scientific research proceeds with the machine-like efficiency that we have come to associate with it (see Kuhn 1962a/1996, 18). Until the scientists in a field establish a normal-scientific research tradition, until they settle on a scientific lexicon, there will be competing schools, each working with its own lexicon. And as long as scientists are divided into competing schools, each generation will begin afresh, reconstructing their field on a new foundation. Little progress can be made in those areas of science that remain in this undeveloped state (15).

But once those working in a scientific field can settle on the fundamentals, they *collectively* become an efficient instrument for developing an elaborate and detailed understanding of the phenomena they study. By employing the same scientific lexicon, they cut nature at the same joints. Further, they learn to discriminate between the phenomena in the same way. They see the world in the same way, even as they see things and differences in the phenomena that others, specifically those outside the group, are unable to see.

Not only are the products of research produced by the group, the infrastructure necessary for successful scientific inquiry is sustained by the group. Kuhn insists that scientific concepts are "the possession of communities" (1991b/2000, 219). Specifically, he wants us to see that scientific concepts "are largely shared by members of a community, [and] their transmission from generation to generation … plays a key role in the process by which the community accredits new members" (219; see also Barnes 1982, 23). This is why Kuhn refers to the cluster of concepts, practices, norms, and standards that relate to the research activities of normal science as a research tradition. Like other traditions, they are sustained and perpetuated only through the concerted efforts of the group.

And the same can be said for the standards of science. Standards of accuracy and reasonableness are both determined and sustained by the research community (see Doppelt 2001; Barnes 2003, 130–32). What counts as evidence or a vindicated prediction is determined by the relevant research community, with different communities tolerating different margins of error. Furthermore, any innovations in standards must, ultimately, be accepted by the relevant research community.

It is worth remembering that Kuhn does not say that the research community is an agent, capable of knowing. He explicitly claims that research communities do not undergo gestalt shifts with a change of theory. Individual scientists may experience something like a gestalt shift, as they move between two competing theories. But the community has no such capacity. There is no group mind. Rather, the community, the collection of similarly trained individuals, sustains the scientific lexicons, exemplars, and standards that enable scientists to pursue their epistemic goals.[5]

Much of what Kuhn says about scientific knowledge being produced by groups is now widely accepted by philosophers. Hence, this part of Kuhn's social epistemology has been integrated into philosophy of science.

SCIENTIFIC CHANGE IS A FORM OF SOCIAL CHANGE

According to Kuhn, significant scientific changes, like changes of theory and the creation of a new scientific specialty, involve significant social changes. Research communities are not static. When the growth

---

[5] There is, currently, some debate about whether the various social groups in science – research teams, specialty communities, and the scientific community as a whole – are aptly described as having the capability to believe something or hold a view (see Gilbert 2000; Wray 2007b; Rolin 2008; Thagard 2010; and Fagan forthcoming). I argue that only research teams have this capability, for they are the only social group in science that has the capacity to accept a view that is not reducible to the views of the individual members of the group (see Wray 2007b). They have this capacity, I claim, in virtue of the fact that they are organized by a functional division of labor. As a result, the members of a research team have an organic solidarity. In contrast, the various members of a scientific specialty have only a mechanical solidarity, that is, they are likeminded.

of knowledge requires it, research communities are capable of significant change. Kuhn notes that scientific research communities are (1) rigid and inflexible when they need to be, and yet (2) capable of adapting to radical changes when they need to, in ways that other social institutions are not. This is part of the essential tension that fascinated Kuhn. Science seems to embody conflicting qualities that are often not seen together in other parts of the social world. This balancing between inflexibility and limberness is what makes science so successful at achieving its goals.

We see this essential tension manifested in Kuhn's characterization of the cyclical pattern of changes that he believes are essential to scientific progress – the change from pre-theoretical scientific research to normal scientific research, the change from normal scientific research to crisis, and the change from crisis to revolution. These changes all involve significant changes in the social order of the research community. A crucial part of the transition from pre-theoretical science to normal science is the emergence of a consensus on a scientific lexicon, that is, agreement about how nature is to be divided into kinds. This requires, to some extent, a loss of flexibility in the community. No longer will the alternative ways of dividing the phenomena that characterize the various schools of thought in the pre-theoretical phase of a scientific field be tolerated. Now, one lexicon or taxonomy will enjoy hegemony in the research community. The suppression of alternative lexicons, the suppression of alternative ways of dividing the phenomena, is essential to establishing a normal-scientific research tradition. But, in time, the accepted lexicon will prove to be an impediment to science, rather than an asset, as it initially was when it enabled the competing schools to finally settle disputes about the fundamentals in the field.

A change in theory, Kuhn claims, requires the erosion of the existing consensus and the emergence of a new consensus (see Kuhn 1962a/1996, 158). This process involves compelling a *community* to work with a new lexicon, to sort the phenomena in fundamentally new ways. Hence, in order for us to understand the nature of theory change what we ultimately need to understand are the causes of the social changes in research communities. In chapter 2, we saw how the Copernican revolution in astronomy was only brought to a close when early accepters of the new theory, including Kepler and Galileo, developed the theory and gathered new data to support it.

And as we saw in chapter 7, the creation of a new scientific specialty also involves a significant social change. Specialty creation, though, differs markedly from theory change, where one theory replaces another.

Specialty formation involves the division of a once-cohesive research community into two separate groups. The process is facilitated by isolation. Only when the scientists in the two groups no longer interact with each other can each group effectively develop the conceptual and instrumental means to realize their now distinct epistemic goals.

The various social changes that Kuhn has identified as part of the process that leads to advances in science have been largely neglected by philosophers. Although many philosophers now understand that scientific knowledge is produced by groups of scientists (see especially Hull 1988; Longino 1990; Kitcher 1993), it is still uncommon to think of scientific change as involving social changes. More precisely, it is still uncommon for philosophers of science to study social changes in science. Hence, as we move forward in developing an epistemology of science, this is a topic that will require more attention. I contribute to this project in the next chapter, when I examine the reception of the theory of plate tectonics in the 1960s.

SOCIAL EPISTEMOLOGY AND THE SOCIOLOGY OF SCIENCE

Given Kuhn's reaction to the work of the Strong Programme, in particular, the effort he exerted to distance his views from their views, one might think Kuhn had little regard for the sociology of science. This is not so. In fact, he believed that sociologists of science were important allies in developing an epistemology of science. Kuhn felt that philosophers of science interested in the epistemology of science will have to work with social scientists in order to effectively answer the questions relevant to their study.

In the past, philosophers of science have generally had little regard for the social dimensions of science. Generally, it was assumed that such things are of no relevance to the *epistemology* of science. Indeed, insofar as the social dimensions of science needed to be studied, it was assumed that sociologists would study them independently of philosophers. Philosophers of science have also generally had little regard for the sociology of science. It has often been assumed, for example, that sociologists have nothing important to say about confirmation and justification, the issues central to the epistemology of science. Many philosophers think that sociologists can contribute only to our understanding of the context of discovery.

This attitude persists. John Worrall even interprets Kuhn's claim that "psychological and sociological factors play ineliminable roles in theory choice" as applying only to the context of discovery, and not the context of justification or appraisal (see Worrall 2003, 97). And some philosophers, for example, Larry Laudan, believe that we should only turn to sociologists for assistance when we seek to account for errors, "where a rational analysis of the acceptance (or rejection) of an idea fails to square with the actual situation" (see Laudan 1977, 202). Laudan thus recommends a clear division of labor where the sociology of science is reduced to the sociology of scientific error. Laudan finds many of the sociological studies of science by the Strong Programme objectionable, like Paul Forman's study of the quick acceptance of the indeterminacy principle in Germany in the 1920s, on the grounds that they try to give a sociological explanation when a rational explanation would be more appropriate (see Laudan 1977, 215–16).

Kuhn recognized the importance of sociology of science for his project early on. In the Postscript to the second edition of *Structure* he explicitly encourages his readers to take note of the important new developments in the sociology of science. He cites, for example, the then cutting-edge work by Warren Hagstrom, Derek Price and Don Beaver, Diana Crane, and Nick Mullins (see Kuhn 1969/1996, 176n. 5). This sociological and historical research was concerned, in one way or another, with determining the community structure of science, including the means of identifying the membership of particular scientific specialties.

Kuhn's call to draw on the sociology of science is another dimension of his social epistemology that has so far been underdeveloped. No doubt, this is partly due to the fact that, with the rise of the Strong Programme, Kuhn was put on the defensive about his relationship to these sociologists of science. The Science Wars have only made matters worse. Philosophers of science are now extremely suspicious of and often quite ignorant about contemporary sociology of science.

Kuhn could have anticipated our current situation, for already at the London conference Lakatos and Popper expressed dismay at the thought that one might get insight into the epistemology of science from sociologists and social psychologists. Lakatos objects to Kuhn's invoking the concept of "crisis" to explain scientific change, arguing that "'crisis' is a psychological concept; it is a contagious panic" (1970, 179). He claims that "*on Kuhn's view there can be no logic, but only psychology of discovery*" (179). And Popper expresses strong doubts about gaining insights from sociologists, arguing that "compared with physics, sociology and psychology are riddled with fashions, and with uncontrolled dogmas" (1970, 57–58). Popper thus makes clear that he believes these fields have no insight to offer us as we seek to understand science and how it works.

In responding to his critics, though, Kuhn makes it clear that sociology is central to his enterprise (see 1970b/2000, 133–34). He insists that psychology cannot answer the questions he poses, for his "unit for purposes of explanation is the normal … scientific group," not the individual scientist (see 133–34).

Because scientific change is a form of social change, the detailed empirical studies of particular episodes of theory change that have been made by sociologists of science are needed to help us better understand the social changes involved in theory change. In particular, empirical studies can help us understand the processes by which a long-accepted theory is abandoned in favor of a new theory. Sociologists are thus useful allies in our efforts to develop an epistemology of science.

It is worth clarifying why Kuhn believes that sociologists of science will need to and can do more than offer insight into the context of discovery. The process by which a theory comes to be accepted in a research community is not aptly characterized as part of the discovery process. Rather, it is more appropriate to characterize it as part of the justification process. Convincing a research community to adopt a new theory often involves a number of tasks, including: (1) generating new data, (2) drawing attention to problems with a long-accepted theory, and (3) convincing others that the problems the new theory can answer are more important than the problems the old theory can answer. These are all matters of justification.

Further, one of Kuhn's key contributions was to show us that it is quite difficult to draw a clear line between a context of discovery and a context of justification. Kuhn's account of scientific discovery, which sees discovery as a drawn-out process, makes this abundantly clear.

The idea of working with scholars in other disciplines, or at least attending to their research findings, is not wholly new to philosophy. In the 1960s W. V. Quine (1969) proposed to naturalize epistemology. For Quine this involves drawing on the best research in psychology as we seek to develop an adequate epistemology. Quine felt such work was indispensable to developing an answer to the question of how we move from the meager sensory input we get from experience to the elaborate theories we have about the world around us, theories that far exceed in content what any single person experiences in a life time.

Many philosophers continue to work on Quine's version of naturalizing epistemology. Indeed, many philosophers working in epistemology keep abreast of developments in psychology and the cognitive sciences in their efforts to answer their research questions. And there has been quite a lot of fruitful research in the philosophy of science that draws on research in the cognitive sciences (see Giere 1988; Thagard 1988, and 1992; Nersessian 2003). In fact, some of it has been inspired by or informed by Kuhn's philosophy of science. Andersen *et al.* (2006), for example, examine Kuhn's theory of concepts in the light of recent research in cognitive science. Kuhn believes that we learn a concept through our experiences of examples of the objects in the class picked out by the concept as well as through experiences of examples of objects in related classes. For example, we learn the concept "goose" through experiences of geese, ducks, and swans, and through learning to discern the similarities and differences between the classes (see Kuhn 1974/1977, 312). Andersen *et al.* (2006) argue that Kuhn's view of concepts and concept learning has been vindicated by research in the cognitive sciences. In particular, research in cognitive science supports Kuhn's claim that people do not have a grasp of the knowledge of the necessary and sufficient conditions for membership in many (if any) of the classes that their concepts pick out (Andersen *et al.* 2006, 6–7; see also Kuhn 1991b/2000, 219). Thus, competency in using concepts does not require such knowledge.[6]

I see Kuhn's proposal that philosophers work with sociologists of science in an effort to develop an epistemology of science as an extension of Quine's project. Quine (1969) suggests that our epistemology needs to be informed by research in psychology. And epistemology has benefited greatly from this influence. Similarly, it seems quite clear that philosophers stand to learn a lot from sociologists and other social scientists as well. Once we accept that science is thoroughly social, we will need to draw on the best contemporary theories of the social world in an effort to understand the social dimensions of science.[7]

---

[6] It is worth noting that in a recent critical review of Andersen *et al.* (2006) Alex Levine argues that "it is startling to find absolutely no engagement with trenchant critiques of Roschian accounts of concepts" like the account of concepts Andersen *et al.* employ (Levine 2010, 375). Because my aim is not to defend Andersen *et al.*'s account of concepts, I will not address the critiques that concern Levine.

[7] In chapters 1 and 2 above, I demonstrated another way in which social scientific research is relevant to the aims of philosophers of science. There, I showed how the literature on political revolution in history and the social sciences could shed light on the nature of theory change, a process Kuhn and others compare to political revolutions. For example, we saw that in both political and scientific revolutions, a crisis is often caused by the articulation of an alternative view. Even though the alternative may have been proposed before there was a crisis, its presentation can help create a crisis. When people, including scientists, see that there are alternatives to the status quo, they begin to think more critically about the status quo and its adequacy. Given that it is a type of social change that philosophers of science seek to understand when they seek to understand theory change, philosophers stand to gain valuable insight from existing social scientific research on

SOME NORMATIVE IMPLICATIONS

So far, most of what I have said about Kuhn's conception of social epistemology has pertained to the descriptive or explanatory project of epistemology. Kuhn's work has provided us with a better understanding of what scientists do and how the culture of science works. But epistemologists also have normative concerns. They want to both *prescribe* how we might do things better, and *evaluate* what scientists do and did in the past. Kuhn also had important insights about these issues. In this section I want to briefly examine some of the normative implications of his social epistemology.

Most importantly, it seems that Kuhn's social epistemology will lead us to evaluate scientists in a manner that is different from the way philosophers have traditionally evaluated scientists. Specifically, we will be concerned to determine how individuals' choices and behaviors affect the prospects of the research community as a whole in their efforts to realize their epistemic goals.

Proponents of the traditional philosophical project, focusing narrowly on individual scientists, require us to look at the evidence a scientist has in our efforts to ascertain whether a particular choice between competing hypotheses was a rational choice. The traditional project involved developing canons of rationality suited to *individual agents*. Kuhn's project, however, involves developing canons of rationality suited to a group.

Kuhn asks us to judge changes in theory from the perspective of the *research community* rather than the perspective of the *individual scientists* involved. In order to understand why the community's perspective is the appropriate perspective in assessing the rationality of scientists, it is useful to examine the change in early modern astronomy from two perspectives, the perspective of the community and the perspective of the individual.

Let us first consider this episode from the traditional individualist perspective. It seems that we are forced to say that either (1) the pioneers – Michael Maestlin, Johannes Kepler, and Galileo Galilei – overstepped the bounds of rationality, accepting the new theory before there was adequate evidence in its support, or (2) the Ptolemaic holdouts were subject to a lapse of rationality, failing to embrace the new theory as the balance of evidence supported it. Assuming that each group had access to roughly the same data, traditional philosophical accounts of

social change. Sociologists, economists, and political scientists can often provide well-developed models of the nature of social groups and social change.

scientific rationality seem incapable of tolerating such different responses to the competing theories. The traditional account presumes a uniform response from scientists working in the same field and from the same body of evidence.

Now, consider this same episode from the perspective of the community. The community of early modern astronomers benefited from the fact that some astronomers, like Maestlin, Kepler, and Galileo, were quick to accept the new theory. These astronomers, though, had to refine and develop the Copernican theory and gather new evidence in its support if they hoped to compel their peers to accept the new theory (see Kuhn 1962a/1996, 156). Galileo even worked on attacking other elements of the accepted Aristotelian theories in physics and hydrostatics in an effort to erode the hold that views associated with Aristotle had on many of his contemporaries. Convincing their peers was the only way that the early converts could bring about a change of theory and ensure a lasting change. But the community of astronomers also benefited from the hold-outs, those reluctant to abandon the Ptolemaic theory, even as the evidence in support of Copernicus' theory mounted. Those astronomers who were initially resistant to the new theory ensured that the research community as a whole was not fickle and inclined to abandon a still promising theory prematurely. Both types of responses to the new theory served a constructive epistemic function. Both types of responses to the new theory can be described as rational or consistent with the canons of scientific rationality. But seeing this requires us to see scientific knowledge as a group product, that is, a product of the concerted efforts of a community of scientists.

This shift in perspective that Kuhn encourages us to adopt will lead us to judge scientists' choices differently and to develop canons of rationality suited to a group. Hence, we are not necessarily committed to judging either the early converts or the holdouts to a change of theory as irrational or overstepping the bounds of rationality. Indeed, as we saw in chapters 1, 2, and 9, and as we will see further in the next chapter, from the perspective of the research community, both groups can and often do serve a *constructive epistemic function*. In this respect, both early converts to a theory and the holdouts aid the community in making the rational choice between competing theories.

Kuhn's conception of scientific rationality is thus quite tolerant. Rational scientists can disagree even as they attend to the same body of data. In this respect, Kuhn's conception of rationality is like Bas van Fraassen's (1989) conception of rationality. For van Fraassen and Kuhn the canons of rationality permit everything except what is explicitly prohibited and thus lead us to regard as rational any choice that is not an explicit violation of the canons of rationality. Alternatively, according to a less tolerant conception of rationality, the canons of rationality prohibit whatever they do not explicitly permit, and thus lead us to regard as irrational anything that is not prescribed by the canons of rationality.[8]

Kuhn believes that subjective factors, that is, factors that are not shared by all members of the research community, often influence scientists during episodes of theory change, especially during the crisis phase when each of the competing theories is regarded as superior by some members of the community (see Kuhn 1977c). It was here, as you will recall from the last chapter, that Kuhn gets mistakenly characterized as an externalist. But one must remember that the function Kuhn attributes to subjective factors in episodes of theory choice is epistemic. A scientist's preference for a simple theory rather than a theory broad in scope, for example, may help ensure that a new, alternative theory is developed and thus given a chance to prove itself, or that a long-accepted theory is not abandoned prematurely. In this way, subjective factors advance the goals of science. As the competing theories are developed by scientists moved by different subjective factors, one theory will emerge as superior, that is, epistemically superior. Consequently, when a consensus does form in a research community, it will be the result of a consideration of epistemic factors. In this respect, Kuhn is clearly an internalist.

The key to seeing the rationality in the process is to take the appropriate perspective, that is, the perspective of the community. From the perspective of the community, the influence of subjective factors serves the important function of dividing the research efforts of those working in the field. This makes the community as a whole more effective at realizing its epistemic goals.

Other philosophers of science have followed Kuhn and sought to deepen our understanding of how the social structure of scientific specialties can affect their ability to achieve their research goals. In different ways, both Philip Kitcher's *Advancement of Science* and David Hull's *Science as a Process* are concerned with the division of labor in science. Kitcher's (1993) concern is to determine what the optimal division of labor

---

[8] Van Fraassen (1989) compares the two forms of rationality, the permissive and the restrictive, to two forms of law, English and Prussian (171–72).

is in cases where there are competing hypotheses. Hull (1988) is concerned with describing some of the different functions served by the different sorts of social groups that constitute a research community. The groups of likeminded research peers who accept the same theories and hypotheses, the groups that Hull calls "demes," often provide assistance in obtaining resources, as well as sympathetic criticism in an effort to ensure that when one presents one's findings publicly, one's presentation is as strong as possible.

There is a second normative implication of Kuhn's account of the social dimensions of science that I want to draw attention to. Research communities not only take care of the training of the next generation, they are also solely responsible for evaluating the contributions of their peers, determining what deserves to be published. Only a specialist has the training required in order to make considered judgments about the quality and merits of scientific research. More than most other social groups in contemporary society scientists are self-policing. This is due, in part, to the fact that scientists are the only ones equipped to evaluate each other. Only those who have internalized the norms, standards, and practices of their field will be able to see the world as their scientific peers see it and thus contribute to the research efforts of the community. And only those scientists will have the requisite background knowledge to evaluate the work of their peers. Scientific knowledge is esoteric knowledge (see Kuhn 1962a/1996, 11 and 24). This fact has long been recognized by scientists. In the Preface to *The Revolutions of the Heavenly Spheres* Copernicus astutely noted that "mathematics is written for mathematicians" (1543/1995, 7). Not only are specialists the only ones who can fully understand scientific contributions, they are also the only ones fit to evaluate them.

In order to develop an adequate epistemology of science, philosophers need to stop thinking of social factors as contaminants or as imposing limitations to our knowing. Clearly, as Barry Barnes (1982) notes, "a socially sustained ordering of the environment" need not be "a socially sustained distortion of it" (23). Following Kuhn, we should see that certain social factors are constitutive of science. They play an integral role in ensuring that scientists are able to realize their epistemic goals. Importantly, as Barnes observes, Kuhn taught us that "the entire framework wherein the reasonable and the social stand in opposition must be discarded" (22).

In closing I want to state what I hope is obvious. Although I have argued that Kuhn believes that social factors can have a profound positive impact on science, I am not claiming that Kuhn believes social factors never have a *negative* impact on science. Indeed, he makes clear that one of the first things a new scientific field needs to do is to develop autonomy, that is, to shield itself from the influence of society. Specifically, Kuhn claims that scientific fields become "mature" when they are able to isolate themselves from the influence of broader social factors. When the sorts of social influences that concern externalist historians of science affect the *outcome* of scientific disputes, then science is in trouble.

CHAPTER II

How does a new theory come to be accepted?

In the previous chapter, I indicated that one of the things we need to understand better is how a new theory comes to displace an older theory. In particular, I indicated that we need to develop a better understanding of the social dimensions of changes of theory. I also suggested that sociological studies of science may be illuminating in advancing our understanding of how the process of theory change unfolds. After all, a change of theory not only involves the *development* of a new scientific lexicon, it also involves the *acceptance* of the new lexicon in the research community. For this to occur, an accepted lexicon must be abandoned.

In this chapter, I want to begin to investigate how a long-accepted theory gets replaced by a new theory in a research community. I want to start, though, by briefly examining some of Kuhn's *speculations* on the process, in particular, his speculations about (1) the role that younger scientists play in the *generation of a new theory*, and (2) the role that older scientists play in the *acceptance of a new theory*. Then, I want to review some evidence that suggests that Kuhn's speculations are mistaken. Finally, I want to examine a particular episode of theory change in the history of science in an effort to develop a better understanding of the nature of the social changes that occur with a change of theory. The object of my study will be the revolutionary change in geology in the 1960s that led to the acceptance of the theory of plate tectonics.

This example suggests that evidence does not affect scientists in the same way, even among those who come to accept the same theory. Some accept a new theory even before the bulk of evidence supports it. Others are led to accept a new theory on the basis of a consideration of new data that support it, responding to the latest developments reported in key articles. Some seem to accept a new theory for other reasons. Finally, it seems that some are reluctant to accept a new theory regardless of the evidence in its favor.

KUHN'S THOUGHTS ON THE SOCIAL DIMENSIONS
OF THEORY CHANGE

In *Structure*, Kuhn makes two interesting claims that are relevant to understanding the process of theory change, and in particular the social dimensions of the process.

First, Kuhn claims that young scientists and those new to a field are more likely to initiate a revolutionary change than are older scientists. According to Kuhn, "almost always the men who achieve [the] fundamental inventions of a new paradigm [that is, a new *theory*] have been either very young or very new to the field whose paradigm they change" (1962a/1996, 90; see also p. 166 there). In one place where Kuhn discusses this claim about young scientists he notes that the "generalization about the role of youth in fundamental scientific research is so common as to be cliché" (90n. 15). But, Kuhn acknowledges that "the generalization badly needs systematic investigation" (90n. 15). Nonetheless, he insists that "a glance at almost any list of fundamental contributions to scientific theory will provide impressionistic confirmation" (90n. 15).

Kuhn claims that there is a good reason why young scientists or outsiders to a field should be the primary sources of novel theories. The creators of new theories, he claims, "are men so young or so new to the crisis-ridden field that practice has committed them less deeply than most of their contemporaries to the world view and rules determined by the old paradigm [that is, the old *theory*]" (1962a/1996, 144). Thus, young scientists are allegedly less entrenched in the status quo, and more capable of seeing the world differently from how they have been habituated to see it during their training and apprenticeship than are their older colleagues. Their older colleagues are alleged to be both less capable and less willing to see the world in ways that conflict with how their early training has taught them to see it. They are less capable because they have been seeing the world through the lenses of the accepted theory longer than young scientists. And they are less willing because their own careers and research contributions are threatened by the prospects of a change of theory.

Second, Kuhn claims that older scientists are especially resistant to changes of theory. Here, Kuhn's concern is not with the development or creation of a new theory, but with its subsequent acceptance by other scientists. Kuhn claims that when a new theory replaces an older theory "some scientists, particularly the older and more experienced ones, may resist [the change] *indefinitely*" (1962a/1996, 152; emphasis added [see also 18–19 and 159]). Kuhn also expresses the complementary of this claim, that is, that young scientists are more accepting of new theories than older scientists (see Kuhn 1969/1996, 203).

In support of this claim about age and resistance to theory change, Kuhn cites both Charles Darwin and Max Planck (1962a/1996, 151). Planck, for example, famously noted that radical scientific changes happen funeral by funeral. That is, often a new theory is only able to displace an older theory after the older generation of scientists dies out (see Planck 1949, 33–34). The claim that older scientists are especially resistant to change has come to be called "Planck's principle" in recognition of the fact that Planck often expressed this bleak view of older scientists. In a footnote, Kuhn also cites a study by Harvey Lehman as support for his claim (see Kuhn 1962a/1996, 90n. 15). But, as Kuhn explains, Lehman's "studies make no attempt to single out contributions that involve *fundamental* reconceptualization" (90n. 15; emphasis added). Rather, Lehman's concerns are with scientific discoveries of any sort.

Importantly, Kuhn does not claim that the resistance of older scientists is an insurmountable barrier to theory change. In fact, he notes that "though a generation is sometimes required to effect the change, scientific communities have again and again been converted to a new paradigm [that is, a new theory]" (1962a/1996, 152).

Further, Kuhn notes that the qualities that make older scientists resistant to theory change, stubbornness and pigheadedness, are the same qualities that make them very effective at the tasks of normal science (see 1962a/1996, 152). In the context of normal scientific research, stubbornness and pigheadedness manifest themselves as epistemic virtues, specifically, as determination and persistence. But Kuhn insists that during episodes of revolutionary science, where a new theory is needed in order for a scientific field to move forward, age is a double liability. It is an impediment to developing radical new theories, and it inclines one to be resistant to theory change.

Kuhn's claims are interesting claims, and *if true* they would have profound implications for understanding the process of theory change. Indeed, claims of this sort have led some philosophers of science to react against Kuhn's view of scientific change. If the acceptance of a new theory is in fact influenced by non-epistemic factors like age, then it seems that evidence does not play the role it should in resolving disputes in science. Clearly, it would be very disconcerting if Planck's principle were true. It would mean that scientists are more dogmatic than they should be, and thus less responsive to the epistemic merits of new data and hypotheses than many assume.[1] In chapter 9, I argued that Kuhn identifies as an internalist and is in fact an internalist. Insofar as subjective factors influence scientists, he believes they merely support the introduction of novelty and ensure an effective division of labor in a research community when more than one theory seems like a viable contender. But with respect to his speculations on the effects of age on theory generation and theory acceptance, he seems like an externalist.

SCRUTINIZING KUHN'S CLAIMS

There is reason to believe that Kuhn is mistaken about both the role that young scientists play in initiating revolutions and the role that older scientists play in preventing or delaying the acceptance of a new theory.

Let us first consider Planck's principle, the claim that older scientists are especially resistant to radical changes. Kuhn alleges that because older scientists have often been key players in developing the currently accepted theories that risk being displaced by a new contender, self-interest makes older scientists especially resistant to innovations.

Hull *et al.* (1978) have subjected Planck's principle to testing. Specifically, they sought to determine whether, during the Darwinian revolution in biology in Britain, younger scientists were more inclined to accept the hypothesis that species evolve, as Planck's principle seems to suggest. They found that the data do not support Planck's principle. Indeed, they did discover that "age is a relevant factor in distinguishing between those scientists who accepted the evolution of species before 1869 and those who did not" (722). But "less than 10 percent of the variation in acceptance is explained by age" (722). Further, contrary to what is implied by Planck's principle, Hull *et al.* found that, "of scientists who accepted the evolution of species before 1869, older scientists were just as quick to change their minds as younger scientists" (722). Consequently, older scientists were not especially resistant to theory change in this case.

Others have also investigated the impact of age on the acceptance of new theories or research programs in other fields, including McCann (1978); Nitecki *et al.* (1978); Diamond (1980); Stewart (1986); Messeri (1988); and Rappa and Debackere (1993). These studies cover a variety of scientific innovations in a range of fields, including geology, chemistry,

---

[1] Many of the post-Kuhnian sociological studies of science seem to assume that scientists are not as responsive to the epistemic merits of new data and hypotheses as philosophers generally assume. See, for example, Bruno Latour's "Give Me a Laboratory and I Will Raise the World" (Latour 1983).

economics, and the study of neural networks. The conclusions reached in these studies vary, but the bulk of evidence supports Hull *et al.*'s conclusion: Planck's principle is a myth. Older scientists are not especially resistant to change.[2]

Let us now consider the claim that younger scientists are especially prone to be the instigators of revolutionary discoveries. Being relatively new to science, young scientists are alleged to be less committed to the status quo than their older colleagues. Consequently, they are more inclined to develop radical new theories or hypotheses (see Lehman 1953; Kuhn 1962a/1996; Gilbert 2000, 45–46). There is plenty of *impressionistic* evidence supporting this claim, as Kuhn suggests. For example, in a study of age and the age structure of scientific research communities, Zuckerman and Merton note that "a long and familiar roster of cases can be provided to illustrate [Kuhn's claim]."

Newton wrote of himself that at 24, when he had begun his work on universal gravitation, and the calculus, and the theory of colors: "I was in the prime of my age for invention, and minded Mathematics and Philosophy more than at any time since." Darwin was 22 at the time of the Beagle voyages and 29 when he formulated the essentials of natural selection. Einstein was 26 in the year of three of his great contributions, among them the special theory of relativity; and finally, eight of the ten physicists generally regarded as having produced quantum physics were under the age of 30 when they made their contributions to that scientific revolution. (Zuckerman and Merton 1973, 513)

Hence, there appears to be some support for this claim about the creative power of young scientists.

But when the issue has been examined more systematically, the data suggest that young scientists are not especially productive of significant discoveries. Harriet Zuckerman (1996), for example, examined the relationship between age and Nobel Prize-winning research. She found that "it is not the young who turn up disproportionately often among those who make prize winning contributions but the middle-aged; 23 percent of the laureates were 40 to 44 years old when they did their prize-winning research but only 14 percent of the run of scientists fall into this

---

[2] Frank Sulloway's (1996) *Born to Rebel* is also concerned with the acceptance of radical ideas, though, unlike these studies, his concern is with the impact of birth order, not age. He found that first-born children tend to be conservative. Further, he found that first-borns are over-represented in the population of scientists.

    Many studies have also examined the claim that younger scientists are *more productive* than older scientists (see, for example, Lehman 1953; Dennis 1956, 1966; Garvey and Tomita 1972; Bayer and Dutton 1977; Stern 1978; Cole 1979; Helmreich *et al.* 1981; Over 1982; Simonton 1984, 1989, 1997; Horner *et al.* 1986; Zuckerman 1996; Kanazawa 2003).

age cohort" (1996, 168). And only 34 percent of the Nobel laureates did their prize-winning research before the age of 35, even though approximately 37 percent of working scientists were under 35 years of age (169). Zuckerman found that the mean age of laureates at the time of their prize-winning research was 38.7, though she also notes that there were differences between fields (166). In physics the average age is 36.1, whereas in physiology and medicine the average age is 40.8.

Zuckerman's study of Nobel laureates suggests that Kuhn is probably mistaken when he claims that young scientists are more likely to develop revolutionary theories. Rather, it seems that young scientists are not especially well positioned to make revolutionary scientific discoveries. If any particular age group is especially well positioned, it appears to be the middle-aged. It is worth noting that Zuckerman's study concerns Nobel Prize-winning research. Such research is clearly important, but it would be a mistake to equate these discoveries with Kuhnian revolutions. After all, as we learned in chapter 1, Kuhn had a very precise conception of what constitutes a scientific revolution. Minimally, it involves a change in the scientific lexicon that results in a regrouping of the phenomena. Much of the research that is honored with a Nobel Prize may not require changes of that sort.[3]

The important point for our purposes is that Kuhn may well be mistaken about the social dimensions of theory change. Older scientists may not be especially resistant to change, and younger scientists may not be the source of the most innovative ideas. Consequently, it would be useful if we could develop a better understanding of the social processes that accompany a change of theory.

THE REVOLUTION IN GEOLOGY

In the remainder of this chapter I want to examine an episode of theory change from the history of science with the aim of developing a better understanding of the dynamics of the social changes that occur when a research community accepts a new theory. I will examine the revolution that occurred in geology in the 1960s that led to the acceptance of the

---

[3] Elsewhere, I have subjected to empirical testing Kuhn's hypothesis about the tendency for young scientists to be the initiators of scientific revolutions (see Wray 2003). I examined the twenty-four scientists responsible for the twenty-five revolutionary discoveries that Kuhn identifies as such in *Structure*. Contrary to what Kuhn suggests, I found that it was the middle-aged scientists who were most inclined to initiate these revolutionary discoveries. Although the sample used in this test is relatively small (N = 28), the examples of revolutionary discoveries are Kuhn's own.

theory of plate tectonics. In addition, I want to draw out the implications of an existing empirical study on this episode of theory change to demonstrate what we can learn from working with social scientists and others doing empirical research on scientific change. In this section, I want to provide a brief account of the nature of the change in theory that occurred. I also want to explain why this instance of theory change is aptly characterized as a Kuhnian revolution.

According to the theory of plate tectonics, the continents lie on large tectonic plates that move across the surface of the earth at a rate of a few centimeters per year. The tectonic plates are pushed apart by a process called seafloor spreading. Lava pushes up through the seams along the edge of the plates, moving the plates and the continents which lie on top of them. Although the process is a very slow process, over many hundreds of thousands of years, significant changes have occurred. South America and Africa, for example, were once joined in one large land mass.[4]

Importantly, the theory of plate tectonics admits of processes that the older theory it replaced did not acknowledge exist. Most significantly, the new theory asserts that many geological processes, including mountain formation, can be explained by the lateral motion of the continents. Tectonic plates had no place in the theory that was accepted before this particular revolution in geology. Indeed, one of the reasons why the theory of continental drift was not widely accepted by geologists earlier in the century when it was first proposed by Alfred Wegener was that geologists had no causal explanation for how the continents could move laterally. Not until the notion of a tectonic plate and the process of seafloor spreading were developed could geologists explain how continents could be moved. The concept "seafloor spreading" was initially proposed in 1960 by Harry Hess (see Frankel 1982, 1; Marshak 2008, 66), and the notion of a "tectonic plate" appears to have been introduced by Tuzo Wilson in 1965 (see Wilson 1965, 343–47; also Glen 1982, 305).[5] Indeed, as Kuhn suggests about scientific discoveries in general, seafloor spreading and tectonic plates were not discovered in one instance. Both of these new concepts,

---

[4] That South America and Africa were once joined is supported by a variety of evidence, including the following. The coastlines of the continents that face each other fit together remarkably well. In addition, the location of past glaciations and the distribution of fossils on both continents strongly suggest that they were once joined (Marshak 2008, 58–61). I thank Paul Tomascak for assistance and information about this episode in the history of geology.

[5] In the 1965 paper in which Wilson suggests that the Earth's surface is divided "into several large rigid plates" (Wilson 1965, 343), he does not call the plates "*tectonic* plates."

especially the latter, developed over time in response to new data and new applications of the new theory.

It is widely recognized that some of the most compelling data in support of the new theory were published in the mid 1960s. This included, importantly, data on magnetic reversals that were gathered from the ocean floor. Two articles in particular, both published in 1966, contain crucial data that show a symmetrical pattern in the magnetic reversals, centered on a ridge between two hypothesized tectonic plates (see Frankel 1982, 31–33). The 1960s was thus a pivotal period in the revolution. Those who believed the continents moved laterally before 1960 were clearly accepting a view that most geologists regarded as unsupported by the available data. And those who accepted the theory of plate tectonics only after 1970 were relatively late in responding to the data.

Geologists and historians of scientists typically describe this change in geology as a scientific revolution (see McArthur and Pestana 1975; Menard 1986, 3; Le Grand 1988, 229; Molnar 2001, 322–23; Marshak 2008, 86; see also Nitecki *et al.* 1978). Indeed, in a paper presented at a conference in 1974, it was already suggested that the revolution in geology was a *Kuhnian* revolution (see McArthur and Pestana 1975). Further, in a widely used college-level geology textbook, the author begins his presentation of the plate tectonics revolution with a brief discussion of Thomas Kuhn's theory of scientific change (see Marshak 2008, 86). Thus, the change of theory in geology that occurred in the 1960s is regarded by geologists as a scientific revolution, and it is regarded as a Kuhnian revolution, although it is doubtful that they have a precise understanding of what is involved in a Kuhnian revolution.[6]

Given that this particular revolution occurred after Kuhn wrote *Structure* it is not surprising that he did not discuss the plate tectonics revolution as an example of a scientific revolution. But this revolution in geology is a classic case of a Kuhnian revolution. It satisfies the three necessary conditions for a scientific revolution outlined in chapter 1.

First, the change that occurred in geology in the 1960s led to the acceptance of a new theory, and thus a new scientific lexicon. Central to this new lexicon was a series of new concepts including the concepts

---

[6] Kuhn's view is misrepresented in some respects in this textbook presentation (see Marshak 2008, 86). For example, Marshak claims that according to Kuhn, after a new theory is proposed, "*almost immediately*, the scientific community scraps the old hypotheses and formulates others consistent with the new paradigm" (86; emphasis added). We saw in chapters 1 and 2 that revolutionary new theories that displace long-accepted theories need not, and often do not, lead to immediate changes.

"continental drift," "seafloor spreading," and "tectonic plate." This change in scientific lexicon involved the replacement of one theory, specifically, a theory that assumes the continents do not move laterally, with a theory that assumes continents can and do move laterally.

Second, there is some evidence that the field of geology was either in a state of crisis or approaching a state of crisis around 1960. The consensus around the long-accepted theory, a fixist or non-mobilist view of the earth, was breaking down. For example, in their study of the acceptance of the theory of plate tectonics, Nitecki *et al.* (1978) report that 22 percent of the geologists in their sample claim to have accepted continental drift by 1960 (661). Moreover, "58% [of the geologists in their sample had] already encountered an advocate of the plate tectonic-continental drift theory prior to 1960" (663). Henry Menard, a participant in the revolution, claims that after World War II "there was no agreement about the most fundamental properties of the earth" among geologists (1986, 3). And Henry Frankel describes the 1960s as "turbulent years for the earth sciences" (1982, 1).

Third, geologists were not able to unequivocally resolve the question of which theory was superior in the early stage of the dispute. Advocates of the competing theories did not agree upon the standards by which the theories should be evaluated. The new theory, with its new concepts, posited the existence and operation of processes that the older theory did not admit. Some of the problems that advocates of the new theory regarded as central to the discipline were largely unacknowledged by advocates of the older theory. Similarly, the new theory was initially unable to explain some phenomena that the previously accepted theory could explain. For example, Naomi Oreskes (2008) claims that the theory of plate tectonics could not explain why the volcanic islands of Hawaii would be "smack dab in the middle of the Pacific plate" (258). According to the theory of plate tectonics, "mountains, volcanoes, and rifts formed at plate boundaries," not in the middle of rigid tectonic plates (258). Further, some geologists doubted that "the planet ... [could] support large-scale crustal movements" (257). Frankel notes additional challenges that the new theory faced, including the lack of a "clear-cut case of a central anomaly over a ridge axis surrounded by the predicted zebra pattern of reversed and normal magnetic anomalies," and the inability of the new theory "to account for ... [changes in] ... wavelength anomalies ... over the ridge flanks" (1982, 23).[7]

---

[7] The term "anomaly" has a very specific meaning in this context. "A magnetic *anomaly* is the difference between the expected strength of the Earth's main field at a certain location and the actual measured strength" (Marshak 2008, 67; emphasis added).

THE GEOLOGICAL REVOLUTION: AN EXAMINATION
OF THE SOCIAL CHANGES

I want now to turn to an analysis of the social changes that occurred during this change of theory in geology. In an effort to develop an understanding of the social changes that occurred, I will draw on an empirical study by Nitecki *et al.* (1978). My aim is to begin the work of uncovering the social processes that enable a new theory to replace a long-accepted theory.

Nitecki *et al.* (1978) examine a sample of geologists with the aim of determining what factors were correlated with the early and late acceptance of the new theory. Their sample included 209 geologists: "128 were Fellows of the Geological Society of America (GSA) and 87 active members of the American Association of Petroleum Geologists (AAPG) (6 were members of both)" (661).

Before proceeding, it is worth noting that, strictly speaking, Nitecki *et al.* (1978) examine the acceptance of the theory of *continental drift* rather than the theory of *plate tectonics*. Continental drift is a theory (or family of theories) that attributes lateral motion to the continents. It is contrasted with a fixist theory, which denies that the continents move laterally. Plate tectonics is a specific version of the theory of continental drift, one that attributes the motion of the continents to the movement of tectonic plates. As a matter of fact, the theory of continental drift was not widely accepted, that is, it was not the dominant theory, until the theory of plate tectonics was developed. The theory of plate tectonics provided a mechanism by which to explain the drifting of continents. It is evident that Nitecki *et al.*'s concern is with the theory of continental drift, rather than specifically the theory of plate tectonics, because they discuss geologists who accept *the theory* before 1960. Tectonic plates, though, were not even hypothesized to exist at that time. Further, as Nitecki *et al.* note, a number of the geologists in their sample did not distinguish between plate tectonics and continental drift, regarding them as synonymous (664).[8]

Nitecki *et al.* distinguish four groups:

(1) Early Acceptors (whom they call "Old Believers"), that is, geologists who adopted the new theory before 1961;[9]

---

[8] A similar situation arose in Hull *et al.*'s (1978) analysis of the acceptance of Darwin's theory of evolution. What they examined was *not* the acceptance of Darwin's theory of evolution by natural selection, but the acceptance of a theory of evolution. Their rationale for doing this was that very few scientists accepted the mechanism of natural selection before 1900, even though many scientists accepted the evolution of species shortly after the publication of *Origin of Species*.

[9] Nitecki *et al.* (1978) report their findings in a way that is somewhat imprecise. At times, when describing the Early Acceptors, they describe them as the geologists who accepted the theory

(2) those who adopted the theory between 1961 and 1970, when most of the new data in support of plate tectonics were collected;
(3) Recent Converts, that is, those who adopted the new theory between 1971 and 1977, the year that Nitecki *et al.* conducted their survey;
(4) Skeptics, those who had not yet accepted the new theory in 1977.

The majority of geologists in Nitecki *et al.*'s sample had accepted the new theory by 1970. Thus, Recent Converts and Skeptics constitute a minority of geologists (662).

Nitecki *et al.* sought to determine if there were important differences between these various groups of geologists. They considered a number of variables, including age, education, country of origin, occupation, specialization, as well as familiarity with the most important publications on plate tectonics. Most of the geologists in Nitecki *et al.*'s sample were born in the United States (1978, 661). They were from a variety of specializations and had a variety of occupations (662). Nitecki *et al.* found that "university teaching is associated with a significantly higher support of plate tectonic theory" (663), and that Early Acceptors were "more likely ... to have a Ph.D. degree, to be university teachers and teach about plate tectonics" (664).

I want now to focus on their findings about familiarity with the most important publications, for this can give us some insight into the role that the research literature plays in bringing about a change of theory. With respect to familiarity with the published literature, each geologist was asked to indicate their familiarity with fifteen key articles which "dealt with arguments for or against plate tectonic-continental drift theory" (662). They were to indicate "whether they were 'unfamiliar with' the publication, 'familiar with' the publication, or had 'read (the publication) in full'" (662). With respect to the five articles that the scientists were most familiar with on the list, 80 percent of the scientists surveyed were either familiar with or read each article, but "only about 25% had read each in full" (662).[10]

Nitecki *et al.* report two interesting patterns with respect to familiarity with the most important publications on plate tectonics. First, Recent Converts and Skeptics were less likely to claim to have read or to claim

of continental drift "*before* 1961" (661), yet at other times, they describe them as the geologists who accepted the theory "*prior to* 1960" (662). This lack of precision, though, does not appear to undermine their results in any significant way.

10 With respect to the remaining ten key articles, only about 45 percent were familiar with or had read each article, and about 10 percent had read each article (see Nitecki *et al.* 1978, 663, table 1).

to be familiar with the fifteen significant publications on plate tectonics. They were thus less familiar with the research literature on plate tecton- ics than the other geologists (664). Second, "there was a tendency for those who had accepted the theory ... before 1960 ... to be [less] familiar with the literature" than those who accepted the theory between 1961 and 1970 (662).[11]

I now want to move beyond Nitecki *et al.*'s findings and draw some conclusions about how the change of theory unfolded. The change of theory occurred in three waves, and each wave was led to accept the theory in a different way. Moreover, there is reason to believe that each wave learned about the evidence in support of the new theory by differ- ent means.

The first wave of scientists, the Early Acceptors, accepted the theory of continental drift even before the bulk of evidence supported it. This group was a minority. As mentioned earlier, according to Nitecki *et al.* "only 22% of the respondents had accepted plate tectonic-continental drift theory prior to 1960" (1978, 662).

It seems that the scientists in the second wave, those who accepted the theory between 1961 and 1970, accepted the theory of continental drift on the basis of the newly gathered evidence. This seems evident from the fact that these geologists were more likely to have either read or be familiar with the fifteen key articles in the field than any other group, including the Early Acceptors (1978, 662).

The third wave, the Recent Converts, consisted of a group of geologists who were less aware of the key literature, in fact, as unaware as geologists who did not accept the theory by 1977.

Left behind by all three waves in the geological revolution was a rela- tively small minority of Skeptics, a group of geologists that constituted 12 percent of the sample under study. These scientists did not change their view even after the vast majority of scientists in their field had accepted the new theory. But these scientists were clearly a minority. Further, some of these scientists *may* have subsequently accepted the new theory before

---

[11] There are additional findings that Nitecki *et al.* (1978) report that are worth noting: "[N]on-U.S. born respondents ... are more familiar with the literature; younger respondents, those with more recent degrees and those who had taken courses in which plate tectonic theory had been discussed, are more familiar with the literature; university teachers are more familiar with the literature than those in other occupations; ... those who attend professional meetings more often are more familiar with the literature, except those who attend more than three meetings a year; and those who publish more ... are more familiar with the literature of plate tectonics, except for those who publish more than three papers a year." (278)

the end of their careers, sometime after 1977, when Nitecki *et al.* conducted their survey.[12]

It is worth noting that the data from this study are consistent with an *internalist* account of the sort that I argued Kuhn favored in chapter 9. There is reason to believe that evidence plays a crucial role in finally bringing a research community around to accepting a new theory. The first and second waves of converts to the new theory were more aware of the relevant research literature than the Recent Converts and the Skeptics. Hence, the *majority* of geologists had already accepted the new theory before geologists less familiar with the key literature accepted the theory. Consequently, it seems that we need not be concerned about external factors leading to consensus formation in science in this case.

One might wonder whether the geologists in the third wave were epistemically negligent, accepting the new theory without adequate knowledge of the evidence supporting the theory or merely because they were under some sort of social pressure from the majority of scientists who had already accepted the new theory. I think it would be presumptuous to conclude that they were epistemically irresponsible. At this point in the revolution many geologists were likely learning about the evidence for the new theory, not from the fifteen key publications, but at second hand, from colleagues and teachers, or as presented in textbooks and other articles that cited these fifteen key articles. The central claims of the new theory, contentious a decade before, were by the 1970s so thoroughly integrated into the field that scientists no longer felt compelled to know, at first hand, the relevant evidence in support of the theory of plate tectonics. These claims were well on their way to becoming "common knowledge."

At one point direct familiarity with the evidence for key claims that are taken for granted in a field is not to be expected. At some point,

---

[12] Unfortunately, given the way that Nitecki *et al.* (1978) report their data, we do not have the means to determine whether the Skeptics were older than those who accepted the new theory. Hence, without further information we do not know whether Kuhn's claim about the death of the older generation is supported by this case. Messeri (1988), though, does examine the age at which geologists accepted the theory of continental drift. Contrary to what is suggested by Planck's principle, he found that the Early Acceptors were generally older. He also found that age had "no effect on middle-period adaptation," that is between 1966 and 1968 (107). There is, though, a problem worth noting with Messeri's study. He claims to be reporting on the views of ninety-six geologists, but his numbers only add to eighty-six (see Messeri 1988, 102). Somewhere, the views of ten geologists were lost.

a discovery or scientific finding becomes part of what Stephen Cole (1992) calls "the core" of a field. Cole claims that "the core consists of a small set of theories, analytic techniques, and facts which represent the given at any particular point in time" (15). Once a claim has achieved this status, the details about when the discovery was first made and what evidence was advanced to support it are of interest to the *historian* of science only.[13] Given the uncritical nature of science education, in particular, the fact that one is instructed in the fundamentals of a field rather than asked to scrutinize them, it should not surprise us to find that scientists are generally unaware of the evidence that first led scientists in their field to accept the theory that has long been accepted.

There is another reason why we need not be concerned by the fact that the geologists who accepted the new theory later were less familiar with the key literature. Recall, from chapter 1, that Kuhn regards the research community as the locus of scientific change. Consequently, a change of theory can still be rational even though not all scientists respond in the same way to the same data. Indeed, as we saw in chapter 9, in order for a research community to be effective, the community must be heterogeneous. The differences between scientists become a valuable source of novelty when a research community is in crisis and the long-accepted theory is deemed inadequate. Hence, rather than signaling a failure of rationality on the part of some geologists, the variety of responses to the new theory indicates that the research community was heterogeneous, as it should be.

This brief examination of the revolution in geology is merely a first step in developing a better understanding of the social dynamics underlying theory change. It is meant to illustrate what sort of data we might seek to gather, and how that data can be used to deepen our understanding of the social dimensions of theory change. If we are to continue to develop an epistemology of science along the lines that Kuhn suggests, this is the type of work we will need to do. And as we see here, philosophers of science can profit from working with sociologists of science and others who are conducting empirical studies of scientific change.

---

[13] Kuhn notes that scientists are often quite ignorant of the history of their field. Generally, a scientist's familiarity with the history of their field is gleaned from the distorted Whig histories that scientists write as part of science textbooks (Kuhn 1962a/1996, 136–37). The purpose of such histories is not, primarily, to reconstruct an accurate account of the past. Rather, according to Kuhn, textbooks are "pedagogic vehicles for the perpetuation of normal science," and the "histories" in them reflect this fact (137).

We can see that scientists do, as a matter of fact, respond to data in different ways. Collectively, a research community is both capable of entertaining novel hypotheses that are not yet supported by the data, and yet not so fickle as to abandon long-accepted hypotheses until sufficient evidence in support of the new hypotheses is gathered.

CHAPTER 12

Where the road has taken us: a synthesis

So where has the road that Kuhn traveled since the publication of *Structure* taken us?

One of my principal aims in this book has been to encourage a re-examination of Kuhn's work. I believe that there are still important insights to gain from his work as we develop an epistemology of science. More precisely, I have argued that: (1) we need to move past the popular negative reading of Kuhn, and (2) in our efforts to understand his constructive contributions to philosophy of science we will benefit from attending to his later work, in particular, Kuhn's mature notion of scientific revolutions and his emphasis on scientific specialization. For the most part, philosophers have seen Kuhn's account of science as a threat to the rationality of science. Consequently, in their discussion of Kuhn's work many philosophers have sought to show either how Kuhn is mistaken in his descriptive account of scientific change, or how he is mistaken about the normative implications of theory change in science. They have seldom sought positive insights from his work.

The proponents of the Strong Programme in the Sociology of Scientific Knowledge have been attracted to the very aspects of Kuhn's view that philosophers have found most unpalatable. They have been intrigued by the apparent threats to the rationality of theory choice and the ambiguities that make resolving scientific disputes difficult. They have also been intrigued by Kuhn's suggestion that researchers working in a normal scientific tradition dogmatically accept the norms, practices, and conceptual framework they were taught. This reaction on their part has only reinforced the common negative attitude toward Kuhn among philosophers. As far as many philosophers are concerned, the proponents of the Strong Programme are merely drawing out the logical consequences of Kuhn's position.

Kuhn's mature view makes the common negative reading of his work unsustainable. One of the main tasks accomplished in Kuhn's later work is a clarification of key concepts for advancing our understanding of science and scientific change, concepts that he made central to philosophy of science. Most important is Kuhn's refined understanding of scientific revolutions. Scientific revolutions are no longer regarded as paradigm changes. The term "paradigm" was elusive and led to many misunderstandings. Revolutions are now classified as taxonomic or lexical changes, changes in the way our theories order the things they aim to account for. Lexical changes play an important role in our efforts to develop a better understanding of the world. But they also pose serious challenges to a research community, as scientists in a community realize that their traditional practices, standards, and concepts are inadequate for modeling the range of phenomena that the community seeks to understand. Scientific revolutions are a distinct class of changes in science that can be distinguished in a principled way from other sorts of changes.

Scientific revolutions are so important in Kuhn's epistemology because they are incompatible with the view that scientific knowledge is cumulative, that scientists are constantly marching ever closer to the truth. The progress scientists make is with respect to saving the phenomena, accounting for observable phenomena.

Given this more precise understanding of scientific revolutions that emerged in Kuhn's later work, it is clear that not all crises in science are resolved in the same way. In addition to changes of theory, which are the outcome of scientific revolutions, research communities sometimes divide, creating a new specialty to deal with the recalcitrant phenomena that caused the crisis in the first place. By creating a new scientific specialty with its own taxonomy or lexicon specially designed to address the hitherto anomalous phenomena, the parent field can continue on, much as before. Importantly, in order to understand the range of phenomena that are of interest to them, scientists must develop a variety of models which are not necessarily consistent with each other. A unified science is thus an unattainable goal. The fragmentation that results from increasing specialization in science is not a temporary stage in the development of science. Rather, it is the means by which scientists achieve their epistemic goals.

According to Kuhn's mature view, a new theory is developed in a field in an effort to account for an anomaly that the accepted theory was unfit to account for. In this respect, competing theories do not address the same problems. And advocates of competing theories do not agree about the relative value of solutions to particular problems in their field. The resulting incommensurability, that is, topic-incommensurability, thus prevents advocates of competing theories from resolving their disputes in a straightforward manner, by appealing to logic and evidence. Because of these differences, it can be challenging for a research community to reach a consensus about which of two competing theories is superior. Incommensurability, however, does not make the rational resolution of such disputes impossible.

Disputes are resolved in a rational manner when the competing theories are refined and new data are collected. The research community is then in a better position to ascertain which theory is superior from an epistemic point of view. Kuhn's critics were led to believe that he thought otherwise, because he seemed to suggest that scientists are often influenced by subjective factors when choosing between competing theories. As a matter of fact, Kuhn did believe that scientists are affected by subjective factors. But rather than being impediments to science, he claims that such factors play a constructive role in times of crisis, ensuring that there is a division of research efforts in the community.

There is a second kind of incommensurability that is also epistemically significant in Kuhn's mature view, the incommensurability that develops between neighboring specialties. This form of incommensurability helps create the barriers that allow the two specialties to develop concepts in ways that serve each specialty's local epistemic needs. Complete and easy communication throughout the scientific community is not always beneficial to science. Rather, we seem to be forced to make a trade-off between easy communication across a wide group of scientists and developing precise concepts that serve relatively small groups of researchers with very specific goals.

I have argued that Kuhn's mature view provides us with an epistemology of science that allows us to develop a richer understanding of both scientific change and scientific knowledge. There are two key aspects of Kuhn's epistemology of science that are crucial to advancing our understanding of science: (1) Kuhn's evolutionary or historical perspective, and (2) Kuhn's social epistemology of science.

Kuhn encourages us to adopt an evolutionary or historical perspective on science. Such a perspective requires us to see science as pushed from behind, like the process of evolution by natural selection. Kuhn encourages us to adopt this perspective because he believes that appeals to the truth really do not explain much at all. Appeals to truth seem especially out of place in explaining the succession of theories in a field, where each successive theory is incompatible with the theory it replaces. There seems to be little reason to think that a series of such changes is a march closer and closer to the true description of reality. This shift to an evolutionary perspective allows us to better understand the dynamics of scientific change. It focuses our attention on the local nature of the goals that drive scientific research. Further, it allows us to see what improvements are made during episodes of theory change. With each change of theory scientists are better able to account for more of the observable phenomena, though sometimes these gains can be made only if scientists divide the domain of their field and thus create a new scientific specialty devoted to the study of recalcitrant phenomena that could not be adequately modeled with the resources of the accepted theory.

Rather than seeing truth as the end of inquiry, Kuhn believes that specialization is the end of inquiry. That is, as scientists seek to realize their epistemic goals, as they seek to develop a better understanding of the natural world, they are led to develop new specialty communities, research communities concerned with modeling a narrower domain. Thus, the increasing accuracy many philosophers traditionally take as evidence that we are converging on the truth is to be explained by the fact that, as science develops, researchers are concerned with a narrower range of phenomena. But as we saw earlier, we are not to think that the increasing accuracy achieved is evidence that we are converging on a true, unified account of the unobservable entities and processes that cause the phenomena, for the various theories developed in different specialties often make claims about the world that are, strictly speaking, incompatible.

I suspect that the shift to a historical perspective will be a difficult transition for philosophers to make. Nonetheless, I believe that Kuhn is correct: an adequate understanding of scientific change and scientific knowledge depends upon us making such a shift. We need to look at science as a process underway, constrained and directed by the accepted theories. Sociologists and historians of science, it seems, have been better at seeing the value in this perspective, for it is now commonplace in these fields. In this respect, Kuhn's position has far greater affinities with sociology of science than with traditional philosophy of science.

Kuhn's epistemology of science is also a social epistemology. First, Kuhn emphasizes the important role that socialization plays in creating scientists. Science education is remarkably uncritical and aims to bring the young aspiring scientist into a research tradition. At the end of the process, the young scientist must see the world as her peers see it, and attend to differences in the phenomena that they deem to be important. Only then can she make a contribution to her field.

Second, Kuhn insists that the locus of scientific change is the research community or specialty. This is a vital contribution and reorientation to understanding scientific change, one that a number of contemporary philosophers of science have made, to some extent. No longer are we able to focus narrowly on how an individual scientist responds to new evidence in our efforts to model or understand theory change. Theory change involves much more than individuals changing their views. A change of theory is a change of view in the community, which involves the replacement of one research tradition by another. Moreover "crisis" describes a state of the research community, not a state of individual scientists. Traditional epistemologies of science that focus on individual scientists and their assessments of theories in light of new data are apt to be blind to these considerations.[1] Consequently, they will miss important aspects of the process of theory change.

But it is important to note that Kuhn's social epistemology does not require us to ascribe mental or psychological properties, like beliefs or perceptions, to the research community. Indeed, Kuhn is adamant that the research community is not an agent. Rather, Kuhn believes that the research community is the locus of change.[2]

Third, on Kuhn's account, theory change is a form of social change. A community must pass through a series of changes in social structure if it is going to abandon a research tradition and replace it with an alternative tradition. Before such a change can take place, the new tradition must be built up, that is, developed. And such development takes time. As we develop our epistemology of science, we need to develop a better understanding of this process. Such a concern is quite different from the sorts of things that have concerned philosophers of science and epistemologists in the past, but these are crucial issues to address as we aim to understand the epistemic dimensions of the social changes in science. For example, we need to identify the means by which consensus is eroded in a research community, as well as the means by which a new consensus is built. Sociologists of science, like Steven Shapin, for example, have been right to insist that the epistemic order in a research community depends on social order. No research community in which the members are deeply divided can effectively advance our knowledge. Now we need to develop a better understanding of how changes of

---

[1] Much work in the epistemology of science is even more abstracted from the social dimensions of science, considering only the logical relations between evidence and theory.

[2] I have explicitly argued elsewhere that research communities, that is, scientific specialties, are not the sorts of groups to which we can ascribe views (see Wray 2007b). Research teams, though, are a different matter. They are capable of holding views not reducible to the views of the constitutive members of the team.

theory can occur without undermining the social order of the research community.

Further, the creation of new specialties plays an important role in enabling scientists to advance their epistemic goals. As we move forward and develop our epistemology of science, it will be important to look with greater care at the process of specialty formation, as well as the effects of specialization on science and scientists. This is because the process of specialty formation offers insight into the increasing accuracy achieved in science, even as changes in theories involve radical changes in our understanding of the basic structure of the world.

Clearly, scientific specialties are not the only social units in science relevant to the epistemology of science. We also need to develop a better understanding of research teams and the groups within scientific specialties from whom scientists seek assistance as they work on their research. David Hull (1988) refers to these groups as demes. Pursuing a better understanding of these groups will take us beyond Kuhn's epistemology, but in the direction that his later research points. This project becomes all the more urgent at a time when most published articles in many scientific fields are multi-authored, some involving teams of twenty or more scientists. In such a research environment, an individual scientist's relationship to her research must be very different from the relationship one has to research that is reported in an article authored by only one scientist.

Fourth, Kuhn's epistemology of science is a social epistemology insofar as it draws on research in the social sciences. If we take seriously his understanding of scientific inquiry and scientific change, then philosophers will inevitably need to work with or at least draw on the work of sociologists of science. In a recent book, Paul Boghossian (2006) argues that those who are influenced by postmodernism and social constructionism, including many sociologists of science, seem to have a fear of knowledge. This fear, he argues, has led them to be critical of any claims to having achieved objectivity.

Boghossian may be correct that many contemporary sociologists of science seem to be afflicted with an unwarranted skepticism. But many mainstream philosophers of science seem to be afflicted with an equally debilitating fear, the fear of the sociology of science. Throughout this book, I have argued that this fear is both ungrounded and unfortunate. It is ungrounded because it is often based on a misreading that is largely due to a misunderstanding of the different objectives that motivate research in sociology of science and philosophy of science. And it is unfortunate because it is only when philosophers and sociologists work together that we will be able to develop an adequate understanding of science and scientific change.

Moreover, if we are going to take the call to naturalize epistemology seriously, in our efforts to naturalize our epistemology of science, we must attend to the work of sociologists of science. The work of sociologists of science can help philosophers to better understand the social structure of research communities and the dynamics of change in such communities. For example, in our efforts to better understand theory change, we will want to study the changes in the social structure of the research community as the consensus on a long-accepted theory is eroded and a new consensus is built around a new competing theory. My brief study of the acceptance of the theory of plate tectonics presented in chapter 11 sheds some light on the structure of scientific communities and the social dimensions of theory change. But more studies of this sort need to be made in our efforts to develop a better understanding of scientific inquiry and scientific knowledge.

In closing, I want to comment on how the social and evolutionary dimensions of Kuhn's epistemology of science fit together. To address this issue, it is useful to compare Kuhn's evolutionary social epistemology of science with Popper's evolutionary epistemology of science. Popper's evolutionary epistemology is not a social epistemology.

Popper believes that scientific theories are subject to a selection process similar to natural selection. By exposing theories to challenging tests, scientists are able to determine which theories to discard. Just as natural selection eliminates unfit variations and species, the selection process in science eliminates unfit theories. The selection is driven by the logic of testing. Theories that entail false predictions are identified through testing and subsequently abandoned. Any progress we make in science toward the truth is a consequence of this process.

Kuhn's evolutionary epistemology is quite different, because in addition to being an evolutionary epistemology, it is a social epistemology. Kuhn denies that logic plays or even could play such a straightforward role in science. As a matter of fact, scientists do not abandon a theory when they observe phenomena that conflict with the theory, as Popper seems to suggest. Rather, as Kuhn notes, such *anomalies* often become the focus of scientists' research efforts.

Further, as noted above, Kuhn does not believe that the success of our current theories is best explained in terms of our getting ever closer to the truth. The successes we have achieved are with respect to predictive accuracy and our ability to account for more observable phenomena. But as far as Kuhn is concerned, these successes can be explained without assuming that we are getting ever closer to a true account of the underlying reality. And the existence of scientific revolutions makes such an assumption untenable. The history of science is a history of theories being replaced by incompatible alternative theories, and there is inadequate continuity in the sequence of theories in a field to warrant accepting a truth-converging explanation.

Kuhn thinks that the success of science has been achieved, in large part, by the social institutions and structure that are constitutive of modern science. That is, the elimination and selection of theories and hypotheses in science is determined, in part, by the social structure of science. The structure of research communities, for example, plays an integral role in scientists' ability to determine which of two competing theories should be retained. Even institutions and practices tied to peer review play a crucial role in the selection process. These social features of science that play a crucial role in the selection of theories constitute the analogue of the environment in the biological world in which the selection of species and variations occur. It is in this way that Kuhn's evolutionary epistemology and social epistemology come together.

It is also worth briefly mentioning the similarities between Kuhn's evolutionary epistemology of science and David Hull's evolutionary social epistemology. Like Kuhn, Hull acknowledges that the social environment in which science is done contributes profoundly to scientists' ability to realize their research goals. Thus, to a large extent, Hull's epistemology of science is consonant with Kuhn's. Indeed, the work of other social epistemologists of science, like Philip Kitcher, Helen Longino, Paul Thagard, and Miriam Solomon, is also consonant, to a greater or lesser extent, with Kuhn's epistemology of science. But Hull's work in the epistemology of science is more empirically oriented than Kuhn's work or the work of these other epistemologists of science. Hull frequently subjects the claims we make about science to empirical testing.[3] He is thus already traveling down the road that Kuhn's work directed us toward. And my own work in the epistemology of science makes me a fellow traveler.

---

[3] For example, in *Science as a Process*, Hull examines how scientists from two competing research programs in zoology – cladists and noncladists – evaluated manuscripts sent to *Systematic Zoology*, a key scientific journal in the field, in an effort to determine if referees are biased by their theoretical allegiance. Hull reports that "the chief difference [he found] is that cladists opted more frequently to both reject and accept manuscripts by their fellow cladists than did noncladists, while noncladists tended to want more extensive modification of the manuscript before publication" (Hull 1988, 333–44).

It is interesting to note, though, that Hull was led to adopt the evolutionary perspective on epistemology by a different path from the one that led Kuhn to adopt it. Recall that Kuhn claims that he adopted this perspective as he worked as a historian, studying science, an institution that is in the process of changing. The evolutionary perspective helped Kuhn realize that it was important to recognize how the previous conditions of the institution limit and shape the direction of the changes undergone. Any future changes are in response to the current conditions. Hull, on the other hand, came to the evolutionary perspective from his work in biology and the philosophy of biology. The dynamics that we see in biological populations seem to be manifested in scientific research communities as scientists respond to new data and other developments in their fields of study.

Much social epistemology is concerned with evaluating the extent to which various social locations are epistemic assets or liabilities. Kuhn's approach to the social epistemology of science is different. It directs our attention elsewhere. Instead of examining how one's social location affects one as an inquirer, he urges us to examine how the social structure of science, the constitutive institutions and practices, enable or hinder scientists, *as a group*, to realize their epistemic goals. One of Kuhn's key insights is that an institution or practice that sometimes aids scientists will, at other times, be an impediment. His distinction between normal science and revolutionary science makes this abundantly clear. Normal science is as effective as it is, ensuring the rapid accumulation of scientific knowledge, because scientists do not question the accepted theory. But a revolutionary change is possible only if one is prepared to abandon the accepted theory. The deep commitment to the accepted theory that scientific training creates thus both aids and impedes scientists in their pursuit of knowledge. Scientific training is effective at creating a uniform community, capable of seeing the same order in the phenomena. But such uniformity needs to be dismantled when a revolutionary change is needed. This is the essential tension that Kuhn talks about.

Clearly, we have come a long way since the publication of *Structure*. Working within the framework of Kuhn's evolutionary social epistemology, we can develop a richer understanding of both scientific change and scientific knowledge.

Bibliography

Abbott, A. 2001. *Chaos of Disciplines*. University of Chicago Press.

Achinstein, P. 2001. "Subjective Views of Kuhn," *Perspectives on Science*, 9:4, 423–32.

Andersen, H. 2001a. *On Kuhn*. Belmont, CA: Wadsworth/Thomson Learning.

2001b. "Kuhn, Conant and Everything: A Full or A Fuller Account," *Philosophy of Science*, 68:2, 258–62.

Andersen, H., P. Barker, and X. Chen. 2006. *The Cognitive Structure of Scientific Revolutions*. Cambridge University Press.

Armitage, A. 1957/2004. *Copernicus and Modern Astronomy*. Mineola, NY: Dover Publications.

Baigrie, B. 1988. "Why Evolutionary Epistemology Is an Endangered Theory," *Social Epistemology*, 2:4, 357–69.

Barker, P. 2002. "Constructing Copernicus," *Perspectives on Science*, 10:2, 208–27.

2001a. "Kuhn, Incommensurability, and Cognitive Science," *Perspectives on Science*, 9:4, 433–62.

2001b. "Incommensurability and Conceptual Change during the Copernican Revolution," in Hoyningen-Huene and Sankey, 241–73.

1999. "Copernicus and the Critics of Ptolemy," *Journal for the History of Astronomy*, 30:4, 343–58.

Barker, P., X. Chen, and H. Andersen. 2003. "Kuhn on Concepts and Categorization," in T. Nickles (ed.), *Thomas Kuhn*. Cambridge University Press, 212–45.

Barnes, B. 2003. "Thomas Kuhn and the Problem of Social Order in Science," in T. Nickles (ed.), *Thomas Kuhn*. Cambridge University Press, 122–41.

1982. *T. S. Kuhn and Social Science*. New York: Columbia University Press.

Barnes, B., and D. Bloor. 1982. "Relativism, Rationalism and the Sociology of Knowledge," in M. Hollis and S. Lukes (eds.), *Rationality and Relativism*. Cambridge, MA: MIT Press, 21–47.

Barnes, B., D. Bloor, and J. Henry. 1996. *Scientific Knowledge: A Sociological Analysis*. University of Chicago Press.

Bayer, A. E., and J. E. Dutton. 1977. "Career Age and Research-Professional Activities of Academic Scientists: Test of Alternative Nonlinear Models and Some Implications for Higher Education Faculty Policies," *Journal of Higher Education*, 48:3, 259–82.

Becher, T., and P. R. Trowler. 2001. *Academic Tribes and Territories*. Buckingham: Society for Research into Higher Education and Open University Press.

Ben-David, J., and R. Collins. 1966/1991. "Social Factors in the Origins of a New Science: The Case of Psychology," in J. Ben-David, *Scientific Growth: Essays on the Social Organization and Ethos of Science*, ed. G. Freudental. Berkeley and Los Angeles: University of California Press, 49–70.

Berlin, I. 2002. "Two Concepts of Liberty," in *Liberty*, ed. H. Hardy. Oxford University Press, 166–217.

Biagioli, M. 1993. *Galileo Courtier: The Practice of Science in the Culture of Absolutism*. University of Chicago Press.

　　1990. "The Anthropology of Incommensurability," *Studies in History and Philosophy of Science*, 21:2, 183–209.

Bird, A. 2003. "Kuhn, Nominalism, and Empiricism," *Philosophy of Science*, 70:4, 690–719.

　　2000. *Thomas Kuhn*. Princeton University Press.

Boghossian, P. 2006. *Fear of Knowledge: Against Relativism and Constructivism*. Oxford University Press.

Bradie, M. 1986. "Assessing Evolutionary Epistemology," *Biology and Philosophy*, 1:4, 401–459.

Brannigan, A. 1981. *The Social Basis of Scientific Discovery*. Cambridge University Press.

Brown, H. 2005. "Incommensurability Reconsidered," *Studies in History and Philosophy of Science*, 36:1, 149–69.

Brown, H. I. 1983. "Incommensurability," *Inquiry*, 26:1, 3–29.

Brown, J. R. 1989. *The Rational and the Social*. London: Routledge.

Bruner, J. 1983. *In Search of Mind: Essays in Autobiography*. New York: Harper and Row.

Bruner, J. S., and L. Postman. 1949. "On the Perception of Incongruity: A Paradigm," *Journal of Personality*, 18:2, 206–23.

Buchwald, J. Z., and G. E. Smith. 2001. "Incommensurability and the Discontinuity of Evidence," *Perspectives on Science*, 9:4, 463–98.

Budd, J. M., M. Sievert, and T. R. Schultz. 1998. "Reasons for Retraction and Citations to the Publications," *Journal of the American Medical Association*, 280:3, 296–97.

Burian, R. M. 2001. "The Dilemma of Case Studies Resolved: The Virtues of Using Case Studies in the History and Philosophy of Science," *Perspectives on Science*, 9:4, 383–404.

Butterfield, H. 1957/1965. *The Origins of Modern Science*, rev. edn. New York: Free Press.

Butts, R. E. 2000. "The Reception of German Scientific Philosophy in North America: 1930–1962," in *Witches, Scientists, Philosophers: Essays and Lectures*, ed. G. Solomon. Dordrecht: Kluwer, 193–204.

Campbell, D. T. 1974. "Evolutionary Epistemology," in P. A. Schilpp (ed.), *The Philosophy of Karl Popper*. La Salle: Open Court, 413–63.

Carnap, R. 1950. "Empiricism, Semantics, and Ontology," *Revue Internationale de Philosophie*, 4:2, 20–40.

Cartwright, N. 1994/1996. "Fundamentalism vs the Patchwork of Laws," in D. Papineau (ed.), *The Philosophy of Science*. Oxford University Press, 314–26.

Cedarbaum, D. G. 1983. "Paradigms," *Studies in History and Philosophy of Science*, 14:3, 173–213.

Chen, X. 1990. "Local Incommensurability and Communicability," *PSA: Proceedings of the Biennial Meeting of the Philosophy of Science Association*, 1 (Contributed Papers), 67–76.

Chen, X. and P. Barker. 2000. "Continuity through Revolutions: A Frame-Based Account of Conceptual Change during Revolutions," *Philosophy of Science*, 67, Supplement, Proceedings of the 1998 Biennial Meetings of the Philosophy of Science Association. Part II: Symposia Papers, S208–S223.

Chubin, D. E. 1976. "The Conceptualization of Scientific Specialties," *Sociological Quarterly*, 17, 448–76.

Cohen, I. B. 1985. *Revolution in Science*. Cambridge, MA: Harvard University Press.

　1974. "History and the Philosophy of Science," in F. Suppe (ed.), *The Structure of Scientific Theories*. Urbana and Chicago: University of Illinois Press, 308–49.

Cohen, L. J. 1973. "Is the Progress of Science Evolutionary?," *British Journal for the Philosophy of Science*, 24:1, 41–61.

Cole, J. R. 1979. *Fair Science: Women in the Scientific Community*. New York: Free Press.

Cole, J. R., and H. Zuckerman. 1975. "The Emergence of a Scientific Specialty: The Self-Exemplifying Case of the Sociology of Science," in L. A. Coser (ed.), *The Idea of Social Structure: Papers in Honor of Robert K. Merton*. New York: Harcourt Brace Jovanovich, 139–74.

Cole, S. 2004. "Merton's Contribution to the Sociology of Science," *Social Studies of Science*, 34:6, 829–44.

　1992. *Making Science: Between Nature and Society*. Cambridge, MA: Harvard University Press.

　1979. "Age and Scientific Performance," *American Journal of Sociology*, 84:4, 958–77.

　1975. "The Growth of Scientific Knowledge: Theories of Deviance as a Case Study," in L. A. Coser (ed.), *The Idea of Social Structure: Papers in Honor of Robert K. Merton*. New York: Harcourt Brace Jovanovich, 175–220.

Collier, J. 1984. "Pragmatic Incommensurability," *PSA: Proceedings of the Biennial Meeting of the Philosophy of Science Association*, 1 (Contributed Papers), 146–53.

Conant, J. B. 1950/1965. "Foreword," in J. B. Conant (ed.), *Robert Boyle's Experiments in Pneumatics*. Cambridge, MA: Harvard University Press, 1–10.

Conant, J., and J. Haugeland. 2000. "Editors' Introduction," in Kuhn 2000a, 1–9.

Copernicus, N. 1543/1995. *On the Revolutions of Heavenly Spheres*, trans. C. G. Wallis. Amherst, NY: Prometheus Books.

Crane, D. 1972. *Invisible Colleges: Diffusion of Knowledge in Scientific Communities*. University of Chicago Press.

Darwin, C. 1878/2003. *On the Origin of the Species by Means of Natural Selection*, corrected 6th edn., ed. J. Carroll. Peterborough, ON: Broadview Press.

Daston, L. 2008. "On Scientific Observation," *Isis*, 99:1, 97–110.

Davies, J. C. 1962. "Toward a Theory of Revolution," *American Sociological Review*, 21:1, 5–19.

Dear, P. 2001. *Revolutionizing the Sciences: European Knowledge and Its Ambitions, 1500–1700*. Princeton University Press.

Demir, I. 2008. "Incommensurabilities in the Work of Thomas Kuhn," *Studies in History and Philosophy of Science*, 39:1, 133–42.

Dennis, W. 1966. "Creative Productivity between the Ages of 20 and 80 Years," *Journal of Gerontology*, 21:1, 1–8.

   1956. "Age and Productivity among Scientists," *Science*, 123:3200, 724–25.

Diamond, A. M. Jr. 1980. "Age and Acceptance of Cliometrics," *Journal of Economic History*, 40:4, 838–41.

Dibner, B. 1980. *Heralds of Science: As Represented by Two Hundred Epochal Books and Pamphlets in the Dibner Library, Smithsonian Institution*, introduction by R. P. Multhauf. Norwalk, CT and Washington, DC: Burndy Library and Smithsonian Institution.

Dobbs, B. J. T. 2000. "Newton as Final Cause and First Mover," in M. Osler (ed.), *Rethinking the Scientific Revolution*. Cambridge University Press, 25–39.

Donovan, A., L. Laudan, and R. Laudan (eds.). 1988. *Scrutinizing Science: Empirical Studies of Scientific Change*. Dordrecht: Kluwer.

Doppelt, G. 2001. "Incommensurability and the Normative Foundations of Scientific Knowledge," in Hoyningen-Huene and Sankey, 159–79.

   1978. "Kuhn's Epistemological Relativism: An Interpretation and Defense," *Inquiry*, 21:1–4, 33–86.

Downes, S. M. 2000. "Truth, Selection and Scientific Inquiry," *Biology and Philosophy*, 15:3, 425–42.

Drake, S. 1970. "The Dispute over Bodies in Water," in *Galileo Studies: Personality, Tradition, and Revolution*. Ann Arbor: University of Michigan Press, 159–76.

Dreyer, J. L. E. 1963. *Tycho Brahe: A Picture of Scientific Life and Work in the Sixteenth Century*. New York: Dover Publications.

Duhem, P. 1917/1996. "Research on the History of Physical Theories," in *Essays in the History and Philosophy of Science*, trans. and ed. R. Ariew and P. Barker. Indianapolis: Hackett Publishing Company, 239–50.

Durkheim, E. 1897/1951. *Suicide: A Study in Sociology*, trans. J. A. Spaulding and G. Simpson. New York: Free Press.

Edge, D. O., and M. J. Mulkay. 1976. *Astronomy Transformed: The Emergence of Radio Astronomy in Britain*. New York: John Wiley and Sons.

Ennis, J. G. 1992. "The Social Organization of Sociological Knowledge: Modeling the Intersection of Specialties," *American Sociological Review*, 57:2, 259–65.

Ereshefsky, M. 1998. "Species Pluralism and Anti-Realism," *Philosophy of Science*, 65:1, 103–20.

1992. "Eliminative Pluralism," *Philosophy of Science*, 59:4, 671–90.

Fagan, B. M. Forthcoming. "Is There Collective Scientific Knowledge?: Arguments from Explanation," *Philosophical Quarterly*.

Farley, J., and G. L. Geison. 1974. "Science, Politics and Spontaneous Generation in Nineteenth-Century France: The Pasteur–Pouchet Debate," *Bulletin of the History of Medicine*, 48:2, 161–98.

Feldman, R. 1988. "Rationality, Reliability, and Natural Selection," *Philosophy of Science*, 55:2, 218–27.

Feyerabend, P. K. 1970/1972. "Consolations for the Specialist," in I. Lakatos and A. Musgrave (eds.), *Criticism and the Growth of Knowledge: Proceedings of the International Colloquium in the Philosophy of Science, London 1965*, vol. IV, reprinted with corrections. Cambridge University Press, 197–230.

Fleck, L. 1935/1979. *Genesis and Development of a Scientific Fact*, ed. T. J. Trenn and R. K. Merton. University of Chicago Press.

Frank, P. 1949. *Modern Science and Its Philosophy*. Cambridge, MA: Harvard University Press.

Frankel, H. 1982. "The Development, Reception, and Acceptance of the Vine-Matthews-Morley Hypothesis," *Historical Studies in the Physical Sciences*, 13:1, 1–39.

Friedman, M. 2003. "Kuhn and Logical Empiricism," in T. Nickles (ed.), *Thomas Kuhn*. Cambridge University Press, 19–44.

2001. *Dynamics of Reason: The 1999 Kant Lectures at Stanford University*. Stanford, CA: CSLI Publications.

2000. "Kant, Kuhn, and the Rationality of Science," *Philosophy of Science*, 69:2, 171–90.

1998. "On the Sociology of Scientific Knowledge and Its Philosophical Agenda," *Studies in History and Philosophy of Science*, 29:2, 239–71.

1974. "Explanation and Scientific Understanding," *Journal of Philosophy*, 71:1, 5–19.

Fuller, S. 2004. *Kuhn vs. Popper: The Struggle for the Soul of Science*. New York: Columbia University Press.

2000. *Thomas Kuhn: A Philosophical History for Our Times*. University of Chicago Press.

1999. *The Governance of Science: Ideology and the Future of the Open Society*. Buckingham: Open University Press.

Galilei, G. 1612/2008. *Discourse on Bodies in Water*, in *The Essential Galileo*, ed. and trans. M. A. Finocchiaro. Indianapolis: Hackett Publishing Company, 85–96.

1612/1960. *Discourse on Bodies in Water*, trans. T. Salusbury, with introduction and notes by Stillman Drake. Urbana: University of Illinois Press.

Galison, P. 1997. *Image and Logic: A Material Culture of Microphysics*. University of Chicago Press.

Garber, D. 2001. "Descartes and the Scientific Revolution: Some Kuhnian Reflections," *Perspectives on Science*, 9:4, 405–22.

Garfield, E. 1980. "Citation Measures of the Influence of Robert K. Merton," in T. Gieryn (ed.), *Science and Social Structure: A Festschrift for Robert K. Merton*. New York Academy of Sciences, 61–74.

Garvey, W. D., and K. Tomita. 1972. "Continuity of Productivity by Scientists in the Years 1968–71," *Science Studies*, 2:4, 379–83.

Gascoigne, R. M. 1987. *A Chronology of the History of Science: 1450–1900*. New York: Garland Publishing.

Gattei, S. 2008. *Thomas Kuhn's "Linguistic Turn" and the Legacy of Logical Empiricism: Incommensurability, Rationality and the Search for Truth*. Aldershot: Ashgate Publishing.

Giere, R. N. 1999. *Science without Laws*. University of Chicago Press.

   1988. *Explaining Science: A Cognitive Approach*. University of Chicago Press.

   1973. "History and Philosophy of Science: Intimate Relationship or Marriage of Convenience?," *British Journal for the Philosophy of Science*, 24:3, 282–97.

Gilbert, M. 2000. "Collective Belief and Scientific Change," in *Sociality and Responsibility: New Essays in Plural Subject Theory*. Lanham: Rowman and Littlefield, 37–49.

Gillispie, C. G. (editor in chief). 1970. *Dictionary of Scientific Biography*. New York: Charles Scribner's Sons.

Gingerich, O. 1975. "'Crisis' versus Aesthetic in the Copernican Revolution," *Vistas in Astronomy*, 17, 85–95.

   1973. "The Role of Erasmus Reinhold and the Prutenic Tables in the Dissemination of Copernican Theory," *Studia Copernicana*, 6, 43–62 and 123–25.

   1971. "The Mercury Theory from Antiquity to Kepler," *Actes du XII Congrès International d'Histoire des Sciences: Paris, 1968*, Paris: Blanchard, vol. IIIA, 57–64.

Glen, W. 1982. *The Road to Jaramillo: Critical Years of the Revolution in Earth Science*. Stanford University Press.

Goldman, A. 1986. *Epistemology and Cognition*. Cambridge, MA: Harvard University Press.

   1999. *Knowledge in a Social World*. Oxford: Clarendon Press.

Goldstone, J. A. 2003. "Comparative Historical Analysis and Knowledge Accumulation in the Study of Revolutions," in J. Mahoney and D. Rueschemeyer (eds.), *Comparative Historical Analysis in the Social Sciences*. Cambridge University Press, 41–90.

   1991. *Revolution and Rebellion in the Early Modern World*. Berkeley and Los Angeles: University of California Press.

Golinski, J. 1998/2005. *Making Natural Knowledge: Constructivism and the History of Science*. University of Chicago Press.

Grantham, T. 1994. "Does Science Have a 'Global Goal?': A Critique of Hull's View of Conceptual Progress," *Biology and Philosophy*, 9:1, 93–94.

Greene, J. C. 1971. "The Kuhnian Paradigm and the Darwinian Revolution in Natural History," in D. H. D. Roller (ed.), *Perspectives in the History of Science and Technology*. Norman: University of Oklahoma Press, 3–25.

Gregory, R. A. 1977. "The Gastrointestinal Hormones: A Historical Review," in Alan L. Hodgkin *et al.*, *The Pursuit of Nature: Informal Essays on the History of Physiology*. Cambridge University Press, 105–32.

Gurr, T. R. 1970. *Why Men Rebel*. Princeton University Press.

Hacking, I. 1999. *The Social Construction of What?* Cambridge, MA: Harvard University Press.

  1993. "Working in a New World: The Taxonomic Solution," in P. Horwich (ed.), *World Changes: Thomas Kuhn and the Nature of Science*. Cambridge, MA: MIT Press, 275–310.

  1983. *Representing and Intervening: Introductory Topics in the Philosophy of Natural Science*. Cambridge University Press.

Hanson, N. R. 1958/1965. *Patterns of Discovery*. Cambridge University Press.

  1961. "The Copernican Disturbance and the Keplerian Revolution," *Journal of the History of Ideas*, 22:2, 169–84.

Harding, S. 1991. *Whose Science? Whose Knowledge?: Thinking from Women's Lives*. Ithaca, NY: Cornell University Press.

  1986. *The Science Question in Feminism*. Ithaca, NY: Cornell University Press.

Hardwig, J. 1991. "The Role of Trust in Knowledge," *Journal of Philosophy*, 88:12, 693–708.

Hayek, F. A. 1960. *The Constitution of Liberty*. University of Chicago Press.

Heidelberger, M. 1976/1980. "Some Intertheoretic Relations between Ptolemean and Copernican Astronomy," in G. Gutting (ed.), *Paradigms and Revolutions: Applications and Appraisals of Thomas Kuhn's Philosophy of Science*. University of Notre Dame Press, 271–83.

Helmreich, R. L., J. T. Spence, and W. L. Thorbecke. 1981. "On the Stability of Productivity and Recognition," *Personality and Social Psychology Bulletin*, 7:3, 516–22.

Henry, J. 2008. *The Scientific Revolution and the Origins of Modern Science*, 3rd edn. Houndmills: Palgrave Macmillan.

Hesse, M. 1976. "Truth and the Growth of Scientific Knowledge," *PSA: Proceedings of the Biennial Meeting of the Philosophy of Science Association*, 2 (Symposia and Invited Papers), 261–80.

Horner, K., J. P. Rushton, and P. A. Vernon. 1986. "The Relation between Aging and Research Productivity of Academic Psychologists," *Psychology of Aging*, 1:4, 319–24.

Hoskin, M. 1997. "Astronomy in Antiquity," in M. Hoskin (ed.), *The Cambridge Illustrated History of Astronomy*. Cambridge University Press, 22–47.

Hoyningen-Huene, P. 2008. "Thomas Kuhn and the Chemical Revolution," *Foundations of Chemistry*, 10:2, 101–15.

  1992. "The Interrelations between Philosophy, History and Sociology of Science in Thomas Kuhn's Theory of Scientific Development," *British Journal for the Philosophy of Science*, 43:4, 487–501.

  1989/1993. *Reconstructing Scientific Revolutions: Thomas S. Kuhn's Philosophy of Science*, trans. A. T. Levine. University of Chicago Press.

Hoyningen-Huene, P., E. Oberheim, and H. Andersen. 1996. "Essay Review: On Incommensurability," *Studies in History and Philosophy of Science*, 27:1, 131–41.

Hoyningen-Huene, P., and H. Sankey (eds.). 2001. *Incommensurability and Related Matters*. Dordrecht: Springer.

Hughes, S. S. 1977. *The Virus: A History of the Concept*. New York: Science History Publications.

Hull, D. L. 2001. *Science and Selection*. Cambridge University Press.

   1988. *Science as a Process: An Evolutionary Account of the Social and Conceptual Developments of Science*. University of Chicago Press.

   1974. "Are the 'Members' of Biological Species 'Similar' to Each Other?," *British Journal for the Philosophy of Science*, 25:4, 332–34.

Hull, D. L., P. D. Tessner, and A. M. Diamond. 1978. "Planck's Principle," *Science*, 202:4369, 717–23.

Kanazawa, S. 2003. "Why Productivity Fades with Age: The Crime–Genius Connection," *Journal of Research in Personality*, 37:4, 257–72.

Kincaid, H. 1996. *Philosophical Foundations of the Social Sciences: Analyzing Controversies in Social Research*. Cambridge University Press.

Kindi, V. 2005. "The Relation of History of Science to Philosophy of Science in *The Structure of Scientific Revolutions* and Kuhn's Later Philosophical Work," *Perspectives on Science*, 13:4, 495–530.

Kitcher, P. 1993. *The Advancement of Science: Science without Legend, Objectivity without Illusion*. Oxford University Press.

Knorr Cetina, K. 1999. *Epistemic Cultures: How the Sciences Make Knowledge*. Cambridge, MA: Harvard University Press.

Koestler, A. 1959. *The Sleepwalkers: A History of Man's Changing Vision of the Universe*. London: Penguin Books.

Kraminick, I. 1972. "Reflections on Revolution: Definition and Explanation in Recent Scholarship," *History and Theory*, 11:1, 26–63.

Kuhn, T. S. 2000a. *The Road since Structure: Philosophical Essays, 1970–1993, with an Autobiographical Interview*, ed. J. Conant and J. Haugeland. University of Chicago Press.

   2000b. "A Discussion with Thomas S. Kuhn," in Kuhn 2000a, 255–323.

   1993/2000. "Afterwords," in Kuhn 2000a, 224–52.

   1992/2000. "The Trouble with the Historical Philosophy of Science," in Kuhn 2000a, 105–20.

   1991a/2000. "The Road since *Structure*," in Kuhn 2000a, 90–104.

   1991b/2000. "The Natural and the Human Sciences," in Kuhn 2000a, 216–23.

   1989/2000. "Possible Worlds in History of Science," in Kuhn 2000a, 58–89.

   1987/2000. "What are Scientific Revolutions?," in Kuhn 2000a, 13–32.

   1983/2000. "Commensurability, Comparability, Communicability," in Kuhn 2000a, 33–57.

   1980. "The Halt and the Blind: Philosophy and History of Science," *British Journal for the Philosophy of Science*, 31:2, 181–92.

1979a. "Foreword," in Fleck 1935/1979, vii–xi.

1979b/2000. "Metaphor in Science," in Kuhn 2000a, 196–207.

1977a. *The Essential Tension: Selected Studies in Scientific Tradition and Change.* University of Chicago Press.

1977b. "Preface," in Kuhn 1977a, ix–xxiii.

1977c. "Objectivity, Value Judgment, and Theory Choice," in Kuhn 1977a, 320–39.

1976/1977. "The Relations between the History and the Philosophy of Science," in Kuhn 1977a, 3–20.

1974/1977. "Second Thoughts on Paradigms," in Kuhn 1977a, 293–319.

1970a/1977. "Logic of Discovery or Psychology of Research?," in Kuhn 1977a, 266–92.

1970b/2000. "Reflections on my Critics," in Kuhn 2000a, 123–75.

1969/1996. "Postscript – 1969," in Kuhn 1962a/1996, 174–210.

1968/1977. "The History of Science," in Kuhn 1977a, 105–26.

1963. "The Function of Dogma in Scientific Research," in A. C. Crombie (ed.), *Scientific Change: Historical Studies in the Intellectual, Social and Technical Conditions for Scientific Discovery and Technical Invention, from Antiquity to the Present.* New York: Basic Books, 347–69.

1962a/1996. *The Structure of Scientific Revolutions*, 3rd edn. University of Chicago Press.

1962b/1977. "The Historical Structure of Scientific Discovery," in Kuhn 1977a, 165–77.

1961/1977. "The Function of Measurement in Modern Physical Science," in Kuhn 1977a, 178–224.

1959/1977. "The Essential Tension: Tradition and Innovation in Scientific Research," in Kuhn 1977a, 225–39.

1957. *The Copernican Revolution: Planetary Astronomy in the Development of Western Thought.* Cambridge, MA: Harvard University Press.

Kuukkanen, J.-M. 2007. "Kuhn, the Correspondence Theory of Truth, and Coherentist Epistemology," *Studies in History and Philosophy of Science*, 38:3, 555–66.

Lakatos, I. 1970/1972. "Falsification and the Methodology of Scientific Research Programmes," in I. Lakatos and A. Musgrave (eds.), *Criticism and the Growth of Knowledge: Proceedings of the International Colloquium in the Philosophy of Science, London 1965*, vol. IV, reprinted with corrections. Cambridge University Press, 91–196.

Latour, B. 1987. *Science in Action: How to Follow Scientists and Engineers through Society.* Cambridge, MA: Harvard University Press.

1983. "Give Me a Laboratory and I Will Raise the World," in K. D. Knorr-Cetina and M. Mulkay (eds.), *Science Observed: Perspectives on the Social Study of Science.* London: Sage, 141–70.

Latour, B., and S. Woolgar. 1986. *Laboratory Life: The Construction of Scientific Facts*, 2nd edn. Princeton University Press.

Laudan, L. 1984. *Science and Values: The Aims of Science and Their Role in Scientific Debate.* Berkeley and Los Angeles: University of California Press.

1977. *Progress and Its Problems: Toward a Theory of Scientific Growth*. Berkeley and Los Angeles: University of California Press.

Laudan, R., L. Laudan, and A. Donovan. 1988. "Testing Theories of Scientific Change," in Donavon *et al.* 1988, 3–44.

Law, J. 1976. "The Development of Specialties in Science: The Case of X-ray Protein Crystallography," in Lemaine *et al.* 1976b, 123–52.

Le Grand, H. E. 1988. *Drifting Continents and Shifting Theories: The Modern Revolution in Geology and Scientific Change*. Cambridge University Press.

Lehman, H. C. 1953. *Age and Achievement*. Princeton University Press.

Lemaine, G., R. MacLeod, M. Mulkay, and P. Weingart. 1976a. "Problems in the Emergence of New Disciplines," in Lemaine *et al.* 1976b, 1–23.

Lemaine, G., R. MacLeod, M. Mulkay, and P. Weingart (eds.). 1976b. *Perspectives on the Emergence of Scientific Disciplines*. Chicago: Aldine.

Levine, A. 2010. "Thomas Kuhn's Cottage," *Perspectives on Science: Historical, Philosophical, Social*, 18:3, 369–77.

Longino, H. E. 2002. *The Fate of Knowledge*. Princeton University Press.

1990. *Science as Social Knowledge: Values and Objectivity in Scientific Inquiry*. Princeton University Press.

McArthur, R. P., and H. R. Pestana. 1975. "Is Continental Drift/Plate Tectonics a Paradigm-Theory?," *Proceedings No. 3: XIVth International Congress of the History of Science*. Tokyo and Kyoto: Science Council of Japan, 105–08.

McCann, H. G. 1978. *Chemistry Transformed: The Paradigmatic Shift from Phlogiston to Oxygen*. Norwood, NJ: Ablex.

McClellan, J. E., and H. Dorn. 1999. *Science and Technology in World History: An Introduction*. Baltimore: Johns Hopkins University Press.

McMullin, E. 1993. "Rationality and Paradigm Change in Science," in P. Horwich (ed.), *World Changes: Thomas Kuhn and the Nature of Science*. Cambridge, MA: MIT Press, 55–78.

Mach, E. 1896/1986. *Principles of the Theory of Heat: Historically and Critically Elucidated*, ed. B. McGuinness. Dordrecht: D. Reidel.

MacIntyre, A. 1981/1984. *After Virtue: A Study in Moral Theory*, 2nd edn. University of Notre Dame Press.

1984. "The Rational and the Social in the History of Science," in J. R. Brown (ed.), *Scientific Rationality: The Sociological Turn*. Dordrecht: D. Reidel, 127–63.

Marshak, S. 2008. *Earth: Portrait of a Planet*, 3rd edn. New York: W. W. Norton and Company.

Marx, W., and L. Bornmann. 2010. "How Accurately Does Thomas Kuhn's Model of Paradigm Change Describe the Transition from the Static View of the Universe to the Big Bang Theory in Cosmology?," *Scientometrics*, 84:2, 441–64.

Masterman, M. 1970/1972. "The Nature of a Paradigm," in I. Lakatos and A. Musgrave (eds.), *Criticism and the Growth of Knowledge: Proceedings of the International Colloquium in the Philosophy of Science, London, 1965*, vol. IV, reprinted with corrections. Cambridge University Press, 59–89.

Mayr, E. 2004. "Do Thomas Kuhn's Scientific Revolutions Take Place?," in *What Makes Biology Unique? Considerations on the Autonomy of a Scientific Discipline*. Cambridge University Press, 159–69.

Menard, H. W. 1986. *The Ocean of Truth: A Personal History of Global Tectonics*. Princeton University Press.

    1971. *Science: Growth and Change*. Cambridge, MA: Harvard University Press.

Merton, R. K. 2004. "Afterword," in R. K. Merton and E. Barber, *The Travels and Adventures of Serendipity*. Princeton University Press, 230–98.

    1996. *On Social Structure and Science*, ed. P. Sztompka. University of Chicago Press.

    1995. "The Thomas Theorem and the Matthew Effect," *Social Forces*, 74:2, 379–422.

    1977. "The Sociology of Science: An Episodic Memoir," in R. K. Merton and J. Gaston (eds.), *The Sociology of Science in Europe*. Carbondale and Edwardsville: Southern Illinois University Press, 3–141.

    1973. *The Sociology of Science: Theoretical and Empirical Investigations*, ed. N. W. Storer. University of Chicago Press.

    1972/1973. "The Perspectives of Insiders and Outsiders," in Merton 1973, 99–136.

    1963/1973. "Multiple Discoveries as Strategic Research Site," in Merton 1973, 371–82.

    1961/1973. "Singletons and Multiples in Science," in Merton 1973, 343–70.

    1959. "Priorities in Scientific Discovery: A Chapter in the Sociology of Science," *American Sociological Review*, 22:6, 635–59.

    1949/1968. "The Self-Fulfilling Prophesy," in R. K. Merton (ed.), *Social Theory and Social Structure*. New York: Free Press, 421–36.

    1945a. "Sociological Theory," *American Journal of Sociology*, 50:6, 462–73.

    1945b. "The Sociology of Knowledge," in G. Gurvitch and W. E. Moore (eds.), *Twentieth-Century Sociology*. New York: Philosophical Library, 366–405.

    1941. "Review of *The Social Role of the Man of Knowledge* by Florian Znaniecki," *American Sociological Review*, 6:1, 111–15.

Merton, R. K., and E. Garfield. 1986. "Foreword," in Price 1963/1986, vii–xiii.

Messeri, P. 1988. "Age Differences in the Reception of New Scientific Theories: The Case of Plate Tectonics Theory," *Social Studies of Science*, 18:1, 91–112.

Mladenović, B. 2007. "'Muckraking in History': The Role of the History of Science in Kuhn's Philosophy," *Perspectives on Science*, 15:3, 261–94.

Molnar, P. 2001. "From Plate Tectonics to Continental Tectonics: An Evolving Perspective of Important Research, from a Graduate Student to an Evolving Curmudgeon," in N. Oreskes (ed.), *Plate Tectonics: An Insider's History of the Modern Theory of the Earth*. Boulder: Westview Press, 288–328.

Mulkay, M. J. 1975 "Three Models of Scientific Development," *Sociological Review*, 23:509–26.

Mulkay, M. J., and D. O. Edge. 1976. "Cognitive, Technical and Social Factors in the Growth of Radio Astronomy," in Lemaine *et al.* 1976b, 153–86.

Mullins, N. C. 1972. "The Development of a Scientific Specialty: The Phage Group and the Origins of Molecular Biology," *Minerva: Review of Science, Learning and Policy*, 10:1, 51–82.

Musgrave, A. E. 1971/1980. "Kuhn's Second Thoughts," in G. Gutting (ed.), *Paradigms and Revolutions: Applications and Appraisals of Thomas Kuhn's Philosophy of Science*. University of Notre Dame Press, 39–53.

Myers, G. 1990. *Writing Biology: Texts in the Social Construction of Scientific Knowledge*. Madison: University of Wisconsin Press.

Nersessian, N. J. 2003. "Kuhn, Conceptual Change, and Cognitive Science," in T. Nickles (ed.), *Thomas Kuhn*. Cambridge University Press, 178–211.

Newton, I. 1726/1999. *The Principia: Mathematical Principles in Natural Philosophy*, trans. I. B. Cohen and A. Whitman, assisted by J. Budenz. Berkeley and Los Angeles: University of California Press.

Newton-Smith, W. H. (ed.). 2000. *A Companion to the Philosophy of Science*. Oxford: Blackwell Publishers.

Nickles, T. 1997. "The Multi-Pass Conception of Scientific Inquiry," *Danish Yearbook of Philosophy*, 32, 11–44.

Nitecki, M. H., J. L. Lemke, H. W. Pullman, and M. E. Johnson. 1978. "Acceptance of Plate Tectonic Theory by Geologists," *Geology*, 6:11, 661–64.

Oberheim, E. 2005. "On the Historical Origins of the Contemporary Notion of Incommensurability: Paul Feyerabend's Assault on Conceptual Conservativism," *Studies in History and Philosophy of Science*, 36:2, 363–90.

Oreskes, N. 2008. "The Devil Is in the (Historical) Details: Continental Drift as a Case of Normatively Appropriate Consensus?," *Perspectives on Science*, 16:3, 253–64.

Over, R. 1982. "Is Age a Good Predictor of Research Productivity?," *Australian Psychologist*, 17:2, 129–39.

Papineau, D. 1996. "Introduction," in D. Papineau (ed.), *The Philosophy of Science*. Oxford University Press, 1–20.

Pedersen, O. 1980. "Tycho Brahe and the Rebirth of Astronomy," *Physica Scripta*, 21:5, 693–701.

Pickering, A. 1984. *Constructing Quarks: A Sociological History of Particle Physics*. Edinburgh University Press.

Pinch, T. J. 1982/1997. "Kuhn – The Conservative and Radical Interpretations: Are Some Mertonians 'Kuhnians' and Some Kuhnians 'Mertonians'?," *Social Studies of Science*, 27:3, 465–82.

  1979. "Paradigm Lost?: A Review Symposium," *Isis*, 70:3, 437–40.

Pinch, T. J., and W. E. Bijker. 1984. "The Social Construction of Facts and Artefacts: Or How the Sociology of Science and the Sociology of Technology Might Benefit Each Other," *Social Studies of Science*, 14:3, 399–441.

Pitt, J. 2001. "The Dilemma of Case Studies: Toward a Heraclitian Philosophy of Science," *Perspectives on Science*, 9:4, 373–82.

Planck, M. 1949. *Scientific Autobiography and Other Papers: With a Memorial Address on Max Planck by Max von Laue*, trans. F. Gaynor. New York: Philosophical Library.

Popper, K. R. 1975/1981. "The Rationality of Scientific Revolutions," in I. Hacking (ed.), *Scientific Revolutions*. Oxford University Press, 80–106.

    1972. *Objective Knowledge: An Evolutionary Approach*. Oxford University Press.

    1970/1972. "Normal Science and Its Dangers," in I. Lakatos and A. Musgrave (eds.), *Criticism and the Growth of Knowledge: Proceedings of the International Colloquium in the Philosophy of Science, London 1965*, vol. iv, reprinted with corrections. Cambridge University Press, 51–58.

    1963. *Conjectures and Refutations: The Growth of Scientific Knowledge*. New York: Harper and Row.

    1959. *The Logic of Scientific Discovery*. New York: Basic Books.

    1946/1950. *The Open Society and Its Enemies*. Princeton University Press.

Price, D. de Solla. 1963/1986. *Little Science, Big Science ... and Beyond*, ed. R. K. Merton and E. Garfield. New York: Columbia University Press.

Quine, W. V. 1969. "Epistemology Naturalized," in *Ontological Relativity and Other Essays*. New York: Columbia University Press, 69–90.

Rand, D. G., A. Dreber, T. Ellingsen, D. Fudenberg, and M. A. Nowak. 2009. "Positive Interactions Promote Public Cooperation," *Science*, 325:5945, 1272–75.

Rappa, M. and K. DeBackere. 1993. "Youth and Scientific Innovation: The Role of Young Scientists in the Development of a New Field," *Minerva: A Review of Science, Learning and Policy*, 31:1, 1–20.

Reichenbach, H. 1938/2006. *Experience and Prediction: An Analysis of the Foundations and the Structure of Knowledge*. University of Notre Dame Press.

Reisch, G. A. 1991. "Did Kuhn Kill Logical Empiricism?," *Philosophy of Science*, 58:2, 264–77.

Renzi, B. G. 2009. "Kuhn's Evolutionary Epistemology and Its Being Undermined by Inadequate Biological Concepts," *Philosophy of Science*, 76:2, 143–59.

Rescher, N. 1978. *Scientific Progress: A Philosophical Essay on the Economics of Research in Natural Science*. University of Pittsburgh Press.

Reydon, T. A. C., and P. Hoyningen-Huene. 2010. "Discussion: Kuhn's Evolutionary Analogy in *The Structure of Scientific Revolutions* and 'The Road since Structure'," *Philosophy of Science*, 77:3, 468–76.

Richardson, A. 2007. "'The Sort of Everyday Image of Logical Positivism': Thomas Kuhn and the Decline of Logical Empiricist Philosophy of Science," in A. Richardson and T. Uebel (eds.), *The Cambridge Companion to Logical Empiricism*. Cambridge University Press, 346–69.

Rolin, K. 2008. "Science as Collective Knowledge," *Cognitive Systems Research*, 9:1–2, 115–24.

Rouse, J. 1987. *Knowledge and Power: Toward a Political Philosophy of Science*. Ithaca, NY: Cornell University Press.

    2003. "Kuhn's Philosophy of Scientific Practice," in T. Nickles (ed.), *Thomas Kuhn*. Cambridge University Press, 101–21.

Sankey, H. 1994. *The Incommensurability Thesis*. Aldershot: Ashgate.

1993. "Kuhn's Changing Conception of Incommensurability," *British Journal for the Philosophy of Science*, 44:4, 759–74.

1991. "Incommensurability, Translation, and Understanding," *Philosophical Quarterly*, 41:165, 414–26.

Sankey, H., and P. Hoyningen-Huene. 2001. "Introduction," in Hoyningen-Huene and Sankey, vii–xxxiv.

Schaffer, S. 1989. "Glass Works: Newton's Prisms and the Uses of Experiment," in D. Gooding, T. Pinch, and S. Schaffer (eds.), *The Uses of Experiment: Studies in the Natural Sciences*. Cambridge University Press, 67–104.

Scheffler, I. 1967. *Science and Subjectivity*. Indianapolis: Bobbs-Merrill.

Schenkman, L. (ed.) 2010. "Random Samples: Who's Bigger?," *Science*, 330:6002, 301.

Schmaus, W. 2008. "A New Way of Thinking about Social Location in Science," *Science & Education*, 17:10, 1127–37.

Schmitt, F. F. 1994. "Socializing Epistemology: An Introduction through Two Sample Issues," in F. F. Schmitt (ed.), *Socializing Epistemology: The Social Dimensions of Knowledge*. Lanham: Rowman and Littlefield, 1–27.

Shapere, D. 1966/1981. "Meaning and Scientific Change," in I. Hacking (ed.), *Scientific Revolutions*. Oxford University Press, 28–59.

1964/1980. "Review of *The Structure of Scientific Revolutions*," in G. Gutting (ed.), *Paradigms and Revolutions: Applications and Appraisals of Thomas Kuhn's Philosophy of Science*. University of Notre Dame Press, 27–38.

Shapin, S. 1996. *The Scientific Revolution*. University of Chicago Press.

1994. *A Social History of Truth: Civility and Science in Seventeenth-Century England*. University of Chicago Press.

1992. "Discipline and Bounding: The History and Sociology of Science as Seen Though the Externalism–Internalism Debate," *History of Science*, 30, 333–69.

1975. "Phrenological Knowledge and the Social Structure of Early Nineteenth-Century Edinburgh," *Annals of Science*, 32:3, 219–43.

Shapin, S. and S. Schaffer. 1985. *Leviathan and the Air-Pump: Hobbes, Boyle, and the Experimental Life*. Princeton University Press.

Shapiro, A. E. 1996. "The Gradual Acceptance of Newton's Theory of Light and Color, 1672–1727," *Perspectives on Science*, 4:1, 59–140.

Sharrock, W., and R. Read. 2002. *Kuhn: Philosopher of Scientific Revolution*. Cambridge: Polity Press.

Simonton, D. K. 2004. *Creativity in Science: Chance, Logic, Genius, and Zeitgeist*. Cambridge University Press.

1997. "Creative Productivity: A Predictive and Explanatory Model of Career Trajectories and Landmarks," *Psychological Review*, 104:1, 66–89.

1992a. "Social Context of Career Success and Course for 2,026 Scientists and Inventors," *Personality and Social Psychology Bulletin*, 18:4, 452–63.

1992b. "Leaders of American Psychology, 1879–1967: Career Development, Creative Output, and Professional Achievement," *Journal of Personality and Social Psychology*, 62:1, 5–17.

1989. "Age and Creativity: Nonlinear Estimation of an Information Processing Model," *International Journal of Aging and Human Development*, 29:1, 23–37.

1984. "Creative Productivity and Age: A Mathematical Model Based on a Two-Step Cognitive Process," *Developmental Review*, 4:1, 97–111.

Sismondo, S. 1996. *Science without Myth: On Constructions, Reality, and Social Knowledge.* Albany: State University of New York Press.

Skocpol, T. 1979. *States and Social Revolutions: A Comparative Analysis of France, Russia, and China.* Cambridge University Press.

Solar, L., H. Sankey, and P. Hoyningen-Huene (eds.). 2008. *Rethinking Scientific Change and Theory Comparison.* Dordrecht: Springer.

Solomon, M. 2001. *Social Empiricism.* Cambridge MA: MIT Press.

Stadler, F. 2007. "The Vienna Circle: Context, Profile, and Development," in A. Richardson and T. Uebel (eds.), *The Cambridge Companion to Logical Empiricism.* Cambridge University Press, 13–40.

Stanford, P. K. 2006. *Exceeding Our Grasp: Science, History, and the Problem of Unconceived Alternatives.* Oxford University Press.

Stern, N. 1978. "Age and Achievement in Mathematics: A Case-Study in the Sociology of Science," *Social Studies of Science*, 8:1, 127–40.

Stewart, J. A. 1986. "Drifting Continents and Colliding Interests: A Quantitative Application of the Interests Perspective," *Social Studies of Science*, 16:2, 261–79.

Stich, S. 1990. *Fragmentation of Reason.* Cambridge, MA: MIT Press.

Sulloway, F. J. 1996. *Born to Rebel: Birth Order, Family Dynamics, and Creative Lives.* New York: Pantheon Books.

Swanson, D. R., and N. R. Smalheiser. 1997. "An Interactive System for Finding Complementary Literatures: A Stimulus to Scientific Discovery," *Artificial Intelligence*, 91:2, 183–203.

Taylor, C. 1985. "Interpretation and the Sciences of Man," in *Philosophy and the Human Sciences: Philosophical Papers 2.* Cambridge University Press, 15–57.

Thagard, P. 2010. "Explaining Economic Crises: Are There Collective Representations?," *Episteme*, 7:3, 266–83.

1999. *How Scientists Explain Disease.* Princeton University Press.

1992. *Conceptual Revolutions.* Princeton University Press.

1988. *Computational Philosophy of Science.* Cambridge, MA: MIT Press.

1980. "Against Evolutionary Epistemology," *PSA: Proceedings of the Biennial Meeting of the Philosophy of Science Association*, 1 (Contributed Papers), 187–96.

Thoren, V. E. 1967. "An Early Instance of Deductive Discovery: Tycho Brahe's Lunar Theory," *Isis*, 58:1, 19–36.

Tilly, C., L. Tilly, and R. Tilly. 1975. *The Rebellious Century: 1830–1930.* Cambridge, MA: Harvard University Press.

Topál, J., G. Gergely, A Erdöhegyi, G. Csibra, and A. Miklósi. 2009. "Differential Sensitivity to Human Communication in Dogs, Wolves, and Human Infants," *Science*, 325:5945, 1269–72.

Toulmin, S. 1972. *Human Understanding*, vol. I. Princeton University Press.

1970/1972. "Does the Distinction between Normal and Revolutionary Science Hold Water?," in I. Lakatos and A. Musgrave (eds.), *Criticism and the Growth of Knowledge: Proceedings of the International Colloquium in the Philosophy of Science, London 1965*, vol. IV, reprinted with corrections. Cambridge University Press, 39–47.

1961. *Foresight and Understanding: An Enquiry into the Aims of Science*. New York: Harper Torchbooks.

Uebel, T. 2008. "Logical Empiricism," in S. Psillos and M. Curd (eds.), *The Routledge Companion to Philosophy of Science*. London: Routledge, 78–90.

Van Fraassen, B. C. 2002. *The Empirical Stance*. New Haven: Yale University Press.

1989. *Laws and Symmetry*. Oxford: Clarendon Press.

1980. *Scientific Image*. Oxford University Press.

Van Helvoort, T. 1994. "History of Virus Research in the Twentieth Century: The Problem of Conceptual Continuity," *History of Science*, 32, 185–235.

Waterson, A. P., and L. Wilkinson. 1978. *An Introduction to the History of Virology*. Cambridge University Press.

Watkins, J. 1970/1972. "Against 'Normal Science'," in I. Lakatos and A. Musgrave (eds.), *Criticism and the Growth of Knowledge: Proceedings of the International Colloquium in the Philosophy of Science, London 1965*, vol. IV, reprinted with corrections. Cambridge University Press, 25–37.

Watson, J. D. 1968. *The Double Helix: Being a Personal Account of the Discovery of the Structure of DNA, a Major Scientific Advance which Led to the Award of a Nobel Prize*. New York: Atheneum.

Weinberg, S. 1998. "The Revolution That Didn't Happen," *New York Review of Books*, 45:15, 48–52.

Westfall, R. S. 2000. "The Scientific Revolution Reasserted," in M. Osler (ed.), *Rethinking the Scientific Revolution*. Cambridge University Press, 41–55.

Westman, R. S. 1994. "Two Cultures or One? A Second Look at Kuhn's *The Copernican Revolution*," *Isis*, 89:1, 79–115.

1975. "The Melanchthon Circle, Rheticus, and the Wittenberg Interpretation of the Copernican Theory," *Isis*, 66:2, 164–93.

White, H. D., B. Wellman, and N. Nazer. 2004. "Does Citation Reflect Social Structure: Longitudinal Evidence from the 'Globenet' Interdisciplinary Research Group," *Journal of the American Society for Information Science and Technology*, 55:2, 111–26.

Williams, L. P. 1970/1972. "Normal Science, Scientific Revolutions and the History of Science," in I. Lakatos and A. Musgrave (eds.), *Criticism and the Growth of Knowledge: Proceedings of the International Colloquium in the Philosophy of Science, London 1965*, vol. IV, reprinted with corrections. Cambridge University Press, 49–50.

Wilson, J. T. 1965. "A New Class of Faults and Their Bearing on Continental Drift," *Nature*, 207:4995, 343–47.

Worboys, M. 1976. "The Emergence of Tropical Medicine: A Study in the Establishment of a Scientific Specialty," in Lemaine *et al.* 1976b, 75–98.

Worrall, J. 2003. "Normal Science and Dogmatism, Paradigms and Progress: Kuhn 'versus' Popper and Lakatos," in T. Nickles (ed.), *Thomas Kuhn*. Cambridge University Press, 65–100.

    1989. "Structural Realism: The Best of Both Worlds?," *Dialectica*, 43:1–2, 99–124.

Wray, K. B. 2010. "Rethinking the Size of Scientific Specialties: Correcting Price's Estimate," *Scientometrics*, 83:2, 471–76.

    2009. "Did Professionalization Afford Better Opportunities for Young Scientists?," *Scientometrics*, 81:3, 757–64.

    2007a. "The Cognitive Dimension of Theory Change in Kuhn's Philosophy of Science," *Studies in History and Philosophy of Science*, 8:3, 610–13.

    2007b. "Who Has Scientific Knowledge?," *Social Epistemology: A Journal of Knowledge, Culture and Policy*, 21:3, 337–47.

    2005. "Does Science Have a Moving Target?," *American Philosophical Quarterly*, 42:1, 47–58.

    2004. "An Examination of the Contributions of Young Scientists in New Fields," *Scientometrics*, 61:1, 117–28.

    2003. "Is Science *Really* a Young Man's Game?," *Social Studies of Science: An International Review of Research in the Social Dimensions of Science and Technology*, 33:1, 137–49.

    2002. "The Epistemic Significance of Collaborative Research," *Philosophy of Science*, 69:1, 150–68.

Zagorin, P. 1973. "Theories of Revolution in Contemporary Historiography," *Political Science Quarterly*, 88:1, 23–52.

Zollman, K. J. S. 2007. "The Communication Structure of Epistemic Communities," *Philosophy of Science*, 74:5, Proceedings of the 2006 Biennial Meeting of the Philosophy of Science Association. Part 1: Contributed Papers, 574–87.

Zuckerman, H. 1996. *Scientific Elite: Nobel Laureates in the United States*. New Brunswick: Transaction.

    1988. "The Sociology of Science," in N. J. Smelser (ed.), *The Handbook of Sociology*. London: Sage, 511–74.

Zuckerman, H., and R. K. Merton. 1973. "Age, Aging, and Age Structure in Science," in Merton 1973, 497–559.

Index

accuracy, 9, 35, 40, 45, 70, 87, 97, 98, 124, 135, 140, 161, 162, 175, 204, 206, 207

Andersen, Hanne, 22, 27, 36, 46, 65, 73, 156, 180

anomaly, 21, 22, 30, 50, 69, 109, 113, 117, 123–24, 128, 129, 135, 158, 202, 207

Aristotelian physics, 32, 41, 74, 99, 160, 182

astronomy, 6, 25, 27, 34–47, 58, 64, 71, 77, 112, 121, 123, 124, 155, 159, 162–63, 176, 181–82

Barker, Peter, 16, 22, 36, 40, 44–45, 46, 180

Barnes, Barry, 93, 114, 135, 156–57, 166–67, 175, 184

Ben-David, Joseph, 119–21, 126

Biagioli, Mario, 107

Bird, Alexander, 22, 23, 24, 31, 32, 49, 63, 65, 71, 85, 104–07, 108, 109, 110, 111, 112, 155, 157

Bloor, David, 114, 156–57, 166–67

Boyle, Robert, 17, 84, 153

Brahe, Tycho, 27, 35, 43, 46–47, 162, 164

Brannigan, Augustine, 132

Brown, Harold, 65, 72, 88

Bruner, Jerome, 50–52, 53, 54

Chubin, Daryl E., 121, 122

Cohen, I. Bernard, 35, 42, 46, 76

Cole, Stephen, 94, 199

Collins, Randall, 119–21, 126

Conant, James B., 51, 54, 56

concept/conceptual change, 3, 8, 9, 21–24, 26–27, 32, 36, 38, 53, 64, 66, 68, 71–73, 75–77, 86, 98, 99, 113–16, 117–22, 125–33, 135–36, 151–53, 156–59, 163–64, 173, 174, 177, 180, 188, 192–94, 201, 202, 203

confirmation, 88–89, 96, 107, 177

consensus, 30, 44–45, 49, 68, 77, 91, 92, 150, 153, 160, 162, 163, 165, 168, 176, 183, 194, 198, 203, 205, 207

constructionism, 5, 10, 94, 145–46, 149–69, 206

convergence, 97, 98, 99, 106, 134, 140, 156, 204, 208

Copernican Revolution, 6, 21, 25, 34–47, 61, 67, 77, 123, 155, 162–63, 176, 182

Copernicus, Nicolaus, 25, 35–38, 40, 43, 44–46, 47, 71, 124, 162, 164, 182, 184

crisis, 10, 15, 17, 21, 24, 45, 74, 123, 135, 139, 159, 168, 176, 178, 183, 187, 194, 199, 202–03, 205

Darwin, Charles, 8, 103, 139, 156, 188, 190

Darwinian revolution, 103, 156, 189

Daston, Lorraine, 111

data, 1, 41, 47, 69, 77, 90, 91, 92–93, 95, 96, 99, 101, 110–11, 113, 114, 153–54, 161, 164, 176, 179, 181, 182, 186, 188, 192–93, 196, 199–200, 203, 205, 209

discovery, 7, 17, 21, 22, 24, 26, 29–30, 31–32, 43, 45, 47, 49, 51, 52, 54, 56, 57, 60, 62, 83, 110, 114–15, 117, 126, 134, 136, 151, 160, 173, 177–79, 190–91, 192, 199

Dobbs, Betty-Jo T., 16, 42

Edge, David O., 121, 122

Einstein, Albert, 76, 99, 160, 190

end of inquiry, 5, 98, 104, 204

endocrinology, 9, 99, 128–29, 131, 136

evidence, 3, 11, 21, 68, 93, 95, 110, 145, 146, 150, 153, 158, 161–62, 175, 181–83, 186, 188, 197–200, 203, 205

evolutionary epistemology, 4, 8, 81–84, 95, 101, 136, 137–40, 208–09

exemplar, 31, 49, 54, 55, 57–61, 63–64, 172–73, 175

experiment, 23, 67, 69, 70, 71, 77, 88–89, 90, 101, 105–11, 114, 116, 128, 129, 134, 153, 164

externalism, 2, 10, 94, 117, 150, 151, 152–54, 160–64, 168, 183, 185, 189, 198

Feyerabend, Paul, 89

finitism, 156–58

Friedman, Michael, 68, 93, 117, 157, 165–66, 167–68, 171

Fuller, Steve, 87, 93, 103, 167, 170

227